WITHDRAWN

Humans, Nature, and Birds

Portrait of Robert Cheseman, 1533, by Hans Holbein the Younger. For a Science Art caption see p. 201.

Humans, Nature, and Birds
Science Art from Cave Walls to Computer Screens

Darryl Wheye and Donald Kennedy
Foreword by Paul R. Ehrlich

Yale University Press   New Haven & London

Published with assistance from the Alfred P. Sloan Foundation's Public Understanding of Science and Technology Program.

Copyright © 2008 by Darryl Wheye and Donald Kennedy.
All rights reserved.
This book may not be reproduced, in whole or in part, including illustrations, in any form (beyond that copying permitted by Sections 107 and 108 of the U.S. Copyright Law and except by reviewers for the public press), without written permission from the publishers.

Designed by James J. Johnson and set in Scala types by BW&A Books, Inc. Printed in the United States of America by R. R. Donnelley, Roanoke, Virginia.

Library of Congress Cataloging-in-Publication Data

Wheye, Darryl.
Humans, nature, and birds : science art from cave walls to computer screens / Darryl Wheye and Donald Kennedy ; foreword by Paul R. Ehrlich.
   p. cm.
Includes bibliographical references.
ISBN 978-0-300-12388-3 (alk. paper)

1. Ornithology. 2. Ornithological illustration. 3. Birds in art. 4. Nature (Aesthetics). 5. Nature—Effect of human beings on. 6. Art and science. I. Kennedy, Donald, 1931– II. Title.
QL673.W48   2008
704.9'4328—dc22   2007050149

A catalogue record for this book is available from the British Library.
The paper in this book meets the guidelines for permanence and durability of the Committee on Production Guidelines for Book Longevity of the Council on Library Resources.

10 9 8 7 6 5 4 3 2 1

Contents

Foreword: Art, Science, and Birds, by Paul R. Ehrlich vii
Preface: What Is Science Art? xiii
Introduction: A Gallery of Science Art xv
Gallery Guide xxi

LOWER GALLERY. Bird Art over the Millennia 1
    ROOM 1. Birds as Icons 5
    ROOM 2. Birds as Resources for Human Use 16
    ROOM 3. Birds as Teaching Tools 29
    ROOM 4. Birds as a Means of Understanding Biology 40
    ROOM 5. Birds as a Means of Promoting Conservation 52

MEZZANINE. Thinking about Aesthetics, the Oldest Bird Paintings, and Painting Nature 64

UPPER GALLERY. How Science and Art Overlap 69
    ROOM 6. Science Art as Its Own Category 72
    ROOM 7. Content, Style, and Medium 95
    ROOM 8. The Importance of Captions 118
    ROOM 9. From Real Public Venues to Virtual Ones 125
    ROOM 10. Science Art, Birds, and Perceptions of Nature 131

APPENDIX 1. Timeline Linking Art, Technology,
   and the Study of Birds   143

APPENDIX 2. A Science Art Checklist for Practitioners   167

Notes   171
Selected Bibliography   185
Acknowledgments   189
Illustration Credits   191
Index   193

Foreword: Art, Science, and Birds

PAUL R. EHRLICH

To some, birds, science, and art may seem like a weird mixture. But they have long been at the center of my intellectual life. Along with friends, wine, and food, they have also been my major focus of enjoyment. So it is with great pleasure that I compose a foreword to a wonderful book on Science Art written by two old, close friends about the relationship between three of four of my favorite loves. The fourth love is my wife, Anne. Anne and I met at the University of Kansas, where she was an art major. I was working on the structural features of butterflies and their evolution, and soon she was working with me, dissecting specimens and drawing features of their minute exterior skeletons, their even tinier hundreds of muscles, their immensely tangled reproductive glands, and the like. Even unartistic scientists often turn to drawing to illustrate points, as I had to when focused on the anatomy of butterflies. That was in the 1950s, but the situation is unchanged today—in many cases, art (or at least illustration, depending on your definitions; I would make no distinction) can better represent science than most photography can. Indeed, we need only compare Ed Wilson's drawings of ants in his scientific work with photographs of ants to see the advantages. (To compare points made with drawings and those made with photographs, see B. Hölldobler and E. O. Wilson, *The Ants* [Cambridge: Harvard University Press, 1990]. To appreciate Wilson's care in illustrating ants see his *Pheidole in the New World: A Dominant, Hyperdiverse Genus* [Cambridge: Harvard University Press, 1993].) The scientist knows what should be visible or obvious, but the camera does not, although a scientist like Tom Eisner who is an artist with a camera can often get equally impressive results. (See T. Eisner, *For Love of Insects* [Cambridge: Harvard University Press, 2003].)

In a sense, the place where Science Art has thrived the longest but, like anatomical drawings of butterflies and ants, mostly away from public view, is in medical illustration. Here again the prevalent idea that somehow photography is unbiased and art is biased is clearly false. The camera cannot deal with such matters as three dimensions, shadowing, and color distortion the way an illustrator can, especially if the illustrator is trained as a scientist or is working collaboratively with one. I would much rather my brain surgeon were guided by a painting supervised by another brain surgeon than guided by a photograph. It is just as impossible to flawlessly "map" reality onto a canvas (or the page of a book) as it is to map it into a human brain, but in a good artist's rendering the critical parts are in clear view and decipherable. All of our perceptions of the world "outside" are filtered from trillions of possible perceptions. All we need to know in order to understand this is that our species, for a couple of hundred thousand years, did not realize that it was bombarded with various kinds of electromagnetic radiation. The radiation was out there, but with the exception of visible light and infrared radiation (heat) we could not detect the radiation until we invented devices that extended our ability to perceive it—radios and Geiger counters, for example. Since then, neurobiologists have abundantly documented how much our sensing of the world, particularly our visual sensing, is a complex interaction involving our environment, our sensory organs, and the brain that interprets the inputs from those organs. It is not so surprising that often Science Art achieves an approximation of what we think is out there better than a photograph does. But Science Art (including photographic Science Art) can do much more than that. It can teach, excite, and aid us, as this wonderful book shows.

The frequent superiority of carefully observed drawings over routine photographs is not just the case with anatomy and small insects. Many, if not most, bird watchers prefer guides with drawings or paintings rather than photographs. Artists' renderings allow birds to be shown in similar poses for the sake of comparison and allows important field marks to be prominently indicated. Field guides of birds provide some of the best examples of Science Art; indeed, Roger Tory Peterson started the whole field-guide business based on Science Art when he used illustrations and indicator lines to show people how to discriminate between similar bird species. Although Peterson's early field guides were quite diagrammatic, later ones were more representational. But I wonder whether some of the most influential ecological science studies, such as Robert MacArthur's brilliant study of niche partitioning in warblers, would have occurred if Peterson had not developed his field guide; MacArthur lay on his back in the woods and documented differences in the behavior of similar warblers that he could identify with ease and accuracy on account of Peterson's work. (See Roger Tory Peterson, *A Field Guide to the Birds* [Boston: Houghton Mifflin, 1934]; Robert H. MacArthur, "Population Ecology of Some Warblers of Northeastern Coniferous Forests," *Ecology* 39 (1958): 599–619.) And how can we ever evaluate the impact and importance of John James Audubon's artistic and scientific landmark *Birds of America* (1830–1838)—whose illustrations made up the first systematic collection of large-format life-size paintings of animals in basically natural poses in natural habitat by a single artist?

Scanning this glorious achievement by Audubon (or the many reproductions of individual plates that are available), we can see what a wise choice Darryl Wheye and Donald Kennedy made in choosing bird art to show the value of Science Art. Years ago Anne and I bought an Audubon print that we could not afford, the Whip-poor-will (*Caprimulgus vociferus*), a member of the nightjar family, because of the biology that it illustrated. It shows three of the birds in front of an oak tree, with one bird about to snatch a big saturniid moth, a cecropia, and another accurately drawn saturniid, an io, in the background. That Audubon portrayal conveyed an important scientific message for its day—that nightjars were nocturnal aerial predators with mouths that could open very wide. (See Plate 61 for a picture of a nightjar.) There was another message, too, for the print took me back to my youth as a moth collector in the northeastern United States. In those days it was easy to get perfect adults of those gorgeous saturniids by finding their cocoons. They have subsequently become scarce because people introduced a parasitic fly to try to control the imported gypsy moth, and the fly has feasted on the saturniids. It would be virtually impossible for a photographer to capture the complex scene that Audubon recorded, and evoke such wonderful memories, but the man who could be thought of as the father of modern avian Science Art presents the scene brilliantly. After Audubon, Science Art devoted to birds thrived—with John Gould in the middle of the nineteenth century, Louis Agassiz Fuertes around the turn of the twentieth century, and a host of other talented artists in more recent years, a number of them formally trained as scientists, George Miksch Sutton, for example (see Plate 67), but many more informally trained as naturalists, alert to the myriad details revealed to the lucky or patient observer of nature. I do not often look at art books, but bird art always attracts my attention and gives me joy. I look forward to receiving catalogues featuring the work of our artist-conservationist friend Tony Angell (see Plates 47 and 48), and I am thrilled by his unsurpassed bird sculptures, a few of which Anne and I are fortunate enough to possess.

Having spent many thousands of hours studying birds in the field, I find that bird art takes me back to special sights—to a Snowy Owl (*Nyctea scandiaca*) standing on the tundra of the Hudson Bay shore, to a Sunbittern (*Eurypyga helias*) dancing down a stream in Costa Rica, to a Peregrine Falcon (*Falco peregrinus*) zooming after ducks in Cornwall. Bird paintings and drawings can depict action that would be very difficult to capture in a photograph and nearly impossible to observe with the naked eye. That is especially true of acts of predation, which in my more than sixty years of field experience are both rarely seen and usually over with too fast to observe with clarity even with binoculars (see Plate 69). Mobbing is a much more common sight, but I have never gotten the same feel for it in the wild as I get from Carl Brenders's magnificent painting (see Plate 53). I often wonder whether the artist who etched the 30,000-year-old owl in Chauvet Cave was not in part just trying to evoke the same sense of awe and mystery that comes over me when I encounter a perched large owl and exchange stares with it (see Plate 2). This example points out one of the attributes of organizing this book as a gallery. The worldwide archive of bird art is immense, but by presenting major topics and arranging a small number of examples sequentially, Wheye and Kennedy make it easy for us to reflect on the long relationship between human beings and birds. Some of the bird images will likely evoke personal memories, adding to the pleasure of this form of bird study. And some of the aspects of science portrayed will likely surprise even longtime students of birds.

As Wheye and Kennedy note, captions can be essential and should be prominently displayed. Despite many visits to the Australian outback I had never been lucky enough to catch a glimpse of the famous mound-building Malleefowl (*Leipoa ocellata*). Soon after November 2005, when I had tried again and failed, Anne discovered a Gould print of the Malleefowl, which we framed. It now occupies a place of honor on the wall of our bedroom. I tried again to spot the bird the following July and failed once more. But in November 2006, in drought-stricken western New South Wales, while searching with colleagues in the vicinity of a known mound nest, I saw that the birds, while building the mound, had scraped about a 50-foot (15-meter) circle around it almost clean of debris. My hopes were raised, even though nothing was moving in the warm sun. Then, out of the corner of my eye, I saw a male Malleefowl, the size of a big rooster, passing by. We were able to watch it climb the mound, scratch the top, and wander around for about twenty minutes. Seeing it was extraordinarily lucky—the record drought and bush fires (perhaps connected with global climate disruption) are threatening many Australian bird populations and were certainly threatening the Malleefowl. Now, whenever I look at the wall across from the foot of our bed, I return to the outback, thanks to John Gould, and recall watching that magnificent and interesting bird.

Part of why I say that the bird is magnificent is that I know its amazing life story. The Malleefowl is a megapod, a member of a group of bird species whose young are superprecocial. Altricial young hatch naked, helpless, and with their eyes closed (robins, finches, eagles, etc.). Precocial young hatch downy and able to walk and be led by parents to food (chickens, ducks, shorebirds). But superprecocial megapod young hatch fully feathered and able to fly within a mere twenty-four hours. That is not the only fascinating thing about megapods. They do not incubate their eggs personally. I have seen one species in the South Pacific that digs tunnels toward hot volcanic streams and lays eggs in those natural incubators. But many megapods are like the Malleefowl, with the male scraping together a huge (often more than 4 feet, or 1 meter, high and 15 feet, or 5 meters, in diameter) pile of organic debris in which the female lays her eggs. The eggs are incubated in cool weather by the heat of decomposition and in hot weather by the direct warmth of the sun. The male monitors the nest temperature (as ours was probably doing) by opening the nest to let heat out if decomposition is making it too hot and later, in cooler weather, by opening and closing the mound morning and afternoon to keep the warmth coming from the sun just right. Think how much more people would appreciate Gould's painting if it had a caption explaining all that, and how many more people would appreciate the Malleefowl if they knew about its distinct, fragile lifestyle.

Birds are, in a sense (as Wheye and Kennedy point out), canaries in our global "mine." For decades the idea of the miners' canaries alerting men underground to the presence of toxic fumes has been the metaphor of choice for the idea of birds alerting humanity to the environmental predicament, just as the idea of springtime without birdsong has been linked to the birth of the modern environmental movement. That movement was started with the publication of Rachel Carson's *Silent Spring* (1962)—a reference to the possibility that the misuse

Mallee Fowl, 1840, by John Gould (1804–1881). Hand-colored lithograph, 14 1/2 × 21 3/4 in. (40.6 × 20.6 cm).

of pesticides would kill so many birds that there no longer would be a wonderful "dawn chorus" in the breeding season. Now birds are threatened by massive habitat destruction and fragmentation, ecological invasions made to serve the human demand for food, fiber, minerals, roads, and living space. Birds face toxification of the planet from pole to pole, increasing rates of climate change, overharvesting for food and for the pet trade, and novel diseases. (For details see my book, co-written with Anne, *One with Nineveh: Politics, Consumption, and the Human Future,* 2nd ed. [Washington, DC: Island Press, 2005].) The decline of species and numbers of birds signals us that we are equally imperiled. I hope that *Humans, Nature, and Birds* will increase the emotive impact of that message by enhancing appreciation of the intertwined scientific and aesthetic value of the world's avifauna. I hope it will add to the chances that future generations can enjoy, as I have, both the feathered creatures in nature and the efforts of artists to depict them—long after I have gone bird watching on Fiddler's Green.

Preface: What Is Science Art?

Works of Science Art skillfully represent truths about the natural world and what it contains, suggesting important connections among people, plants, animals, and their environment and teaching us indirectly about nature itself.

We have sometimes been asked: "Why 'Science Art'?" The term neither describes the science of artistic creation nor merely depicts scientific events or laboratory experiments. Instead, we see in Science Art a kind of fusion in which a painting or sculpture or photograph or other piece of art says something about the natural world and how it works. Whether or not the artist is motivated by a scientific purpose, a work of Science Art can enrich the viewer with a sense that its subject is connected with, and could help explain, relationships. The artist sometimes uses scientific knowledge and findings; sometimes these emerge only because the artist's execution is sensitive and faithful to these relationships. To qualify as Science Art and to work well, the rendering should be accompanied by an explanatory caption that helps the viewer decode the underlying science.

We have also sometimes been asked: "Why not 'Environmental Art' or 'Wildlife Art' or 'Nature Art'?" Works of art that represent truths about the world and what it contains certainly include these categories, although not all examples of Environmental Art, Wildlife Art, and Nature Art convey truths. Here, too, it is the explanatory caption that provides the viewer with access to truths that might otherwise be overlooked.

Introduction: A Gallery of Science Art

Art often devotes itself to exploring our relationship with nature. Narratives about that relationship have been recorded in landscapes, whether or not our ecological footprint is evident in the picture and irrespective of locality, be it urban, pastoral, or marine. Artists sometimes provide entry into these views of nature through the use of particular representatives. More frequently than we realize, and for good reason, these representatives are birds. We like their color and quickness and admire to the point of envy their mastery of the air. In nature, a mere speck of brilliant plumage as a bird perches quietly or the flick of white wing bars as it darts from branch to branch can catch the eye. On canvas, even when the bird is presented independently of its native environment, we are drawn to the sheen of a spread wing with its wing bars now as discernible as the stripes on a sergeant's sleeve. We are a curious species, wanting to understand behaviors and relationships. In the wild, a bird's voice is often our first indication of its presence, leading us to seek the owner and to determine what it might be communicating and to whom. On canvas, a bird directing strenuous calls leads our eye to the source of its concern. Glimpses of dramatic group behaviors, like the mating migrations of Emperor Penguins in the hostile environment of the deep Antarctic, let us appreciate the ability of life to extend into every corner of the globe, whereas human behaviors, like the global efforts to contain the spread of the H5N1 bird virus, give us new glimpses of the extent to which what affects birds also affects us.

Birds have thus been an artistic focus not only because they are lively, appealing, colorful, and dramatic but also because their behavior is interesting, and they are important to our well-being. The focus dates back to the Paleolithic period, when early artists conveyed messages about animals, including birds and their environment, and continues to this day. Decoding the images, however, can pose a challenge, for when they lack captions—as most do—we often rely on slim clues to guide us. At least with cave wall art, the images are still in situ. Even the oldest bird image yet found—the 30,000-year-old etching in Chauvet Cave that, at first glance, is only suggestive of a generic owl—turns out to be remarkably accurate, and its presence provides clues about its value to the local people, which makes it particularly interesting.

Today there is growing literary interest in the history of nature and human attitudes toward it, in large part because of the escalating concerns about uses of the environment and managing it in a sustainable way. Serious interest in nature—in short supply when the natural environment was seen mainly as a source of physical resources—has expanded especially rapidly over recent decades. Artistic expression of the relationship between people and their natural environment continues to enhance and broaden that interest—in part because images that include the science underlying the subject portrayed span educational, cultural, and language differences.

This book is for those who seek a keener awareness of what is taking place in the natural world and how artists have recorded it. We will show how these narratives can be presented and looked at in a way that encourages insights about nature and how it works. Specifically, we

Owl in Chauvet Cave (see Plate 2)

especially in the realm of nature—as subject matter. The interplay of science and art has thus had a long, continuous history. On the one hand, many artists (even including the one who etched the Chauvet owl more than 30,000 years ago; see Plate 2) seem to have a deep understanding of their environment and its organisms and often present behaviors and events that science will later describe in words. On the other hand, as scientific understanding about evolution, ecology, behavior, and conservation has grown, artists have incorporated these new findings into their work. Most art historians have paid relatively little attention to this dimension of art. We will explore—through the captions—that element and underscore what the artworks seem to be conveying about nature, about the ecological role of the subject, and about human relationships with nature. The messages will sometimes involve problems associated with habitat disturbance and loss, or ecological imbalance—and at other times they may lift the viewer to an inspiring vision of undisturbed nature.

will use science to help expose messages within the art. Discussion of these messages is provided in one-page captions facing each plate. By keeping the captions short and forgoing lengthy discussions of art history, we hope readers will find it easy to absorb the information in the captions as they view the pictures.

Our approach requires a term for artwork with scientific content or a scientifically based message. "Science Art" seems the most appropriate because it narrows the space between science and works that portray nature in a biologically valid way. It is essential that Science Art include captions that can increase the viewer's appreciation of the science embedded in the art. Our hope is that combining captions with art will raise the visibility of scientifically valid art about nature, and will encourage more artists to produce it, more teachers to use it, and more viewers to seek it.

Scientists often use art the same way they use other essential technologies; and art has often used science or scientific phenomena—

We have selected sixty images to frame our discussion. All but one features birds, and most were gleaned from Darryl Wheye's yearlong exploration of the massive 30,000-year pictorial archive. This legacy of 1,500 or so generations of artists is divided among more than 25,000 museums, 18,000 major galleries, a vast array of original sites (for example, caves, rock formations, the walls of churches and other institutions), and the walls and vaults of countless collectors.[1] Initially, she selected 500 images featuring birds and sorted them by date and content. As general patterns emerged, it became obvious that images illustrating the use of birds as icons, as resources useful to people, and as teaching tools could be traced all the way back to cave walls. So could scientifically accurate images of bird biology and images that convey information useful for current conservation efforts.

In the early 1990s, Wheye used the results of that survey to develop a graphic database and annotate a timeline charting the development of ornithology as a science. The database cov-

Modified Owl in Chauvet Cave (see Plate 3)

ered 100 topics, and the assemblage became a 1,800-page experimental Web site that was tested in Paul Ehrlich's undergraduate course on bird biology at Stanford University, with which Donald Kennedy is associated. In 2001, we assessed the utility of the database for an introductory course that Kennedy was co-teaching on the study of nature. Although online image use fees proved beyond the project's budget at that point, we have produced a Web site related to this book (scienceart-nature.org).

In 2002, we took another approach to raising the visibility of biologically valid images of birds. Again in association with Paul Ehrlich, we launched an online registry at Stanford University that features the work of more than 100 bird artists (birds.stanford.edu). Nine ornithological and birding organizations, including the National Audubon Society and the Royal Society for the Protection of Birds, three natural history and art museums, including the Smithsonian Institution, two academic centers, and a wildlife artist organization have endorsed the registry. In 2008, the registry initiated an online exhibit of Science Art produced by participants—yet another manifestation of how the digital age is improving access to the scientific content embedded in art. *Humans, Nature, and Birds* includes the works of a number of registry participants because these images demonstrate certain distinctive aspects of this approach to art.

The book's "gallery" of sixty images has two levels—lower and upper—each with five rooms. It may be helpful to begin with a few comments about the exhibits.

Gleaned primarily from the original sample of 500, the images in the Lower Gallery provide historical perspective. Each room is dedicated to one of the five categories discerned in the thousands of images examined: birds as icons; birds as resources used by people; birds as teaching tools; "illustrations" of bird biology; and "illustrations" useful for conservation issues. The first image in the first room is the Chauvet owl, chosen because it is the oldest wall image of a bird found to date, suggesting that owls are among the oldest bird icons.

The caption accompanying that example—like the captions that follow—provides information on the probable relationship of the bird to human society at the time the art was produced and the importance of the image as an example of Science Art, including information about nature or conservation that it portrays or suggests. Because many of the early images are unclear or difficult to recognize in terms of modern avian taxonomy, and because the narratives are puzzling, we have done some detective work—often presenting interpretations different from the standard interpretations given by curators or historians. We make no claim of certainty in decoding the science; on the contrary, this exercise, which we find fascinating and potentially useful, is open to all. We have divided the caption for each plate into two sections. One section provides general information about the art, the artist, the era, the medium, and so forth, and the other gives general information about the plate's scientific content.

When we say that a work of art fits the category of Science Art, we are not suggesting that every artist whose work is shown here deliber-

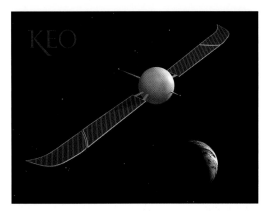

KEO *the Winged Satellite*, 1998, by Jean-Marc Philippe

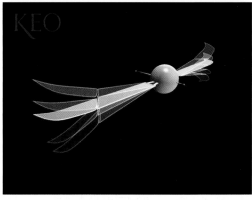

KEO *the Archeological Bird of the Future*, 1998, by Jean-Marc Philippe

ately portrayed an aspect of science or selected a narrative with a message that is scientifically valid—although many certainly have. Viewers and artists alike approach art with experiences and agendas based on their histories, so there are countless avenues of creativity and numerous ways to enter into the resulting works of art. What is different here is that science provides a framework for entry that is independent of knowledge about the artist or the style or the culture—although familiarity with any of those things may provide a framework for understanding the work.

Since the days of the cave artist, Science Art featuring birds has appeared and reappeared. It is presently flourishing. One contemporary example is "The Nest," the 80,000-seat Beijing stadium built for the 2008 Summer Olympics. Another example is an avian-like icon that UNESCO chose as the "Project of the XXIst Century." KEO, a passive satellite scheduled for launch soon (contingent on the successful launch of the Ariane 5 rocket) and return in 50,000 years, will carry messages and a time capsule containing an engraved diamond. The engraving provides the code for human DNA, and the diamond encapsulates a drop of human blood and samples of air, seawater, and soil. Like the image of the massive nest in Beijing, whose steel beams are bent and twisted to resemble woven sticks, the image of KEO in orbit above Earth provides an interesting metaphor and a melding of science and art. KEO's albatross-like wings were added so that the satellite would "appeal to the human imagination as a mythical figure of a bird as the messenger of the people of the twenty-first century." The temperature-sensitive metal wings will flutter slightly, like a soaring bird's. They will flex downward each time the satellite enters Earth's shadow and flex upward as it returns to direct sunlight.[2]

After the Lower Gallery, the Mezzanine presents ideas about aesthetics along with brief discussions about cave art and painting nature, with particular reference to the art historian Sir Ernest Gombrich. This presentation leads to the Upper Gallery, which gives an overview of artistic aspects of Science Art. About one-quarter of these images were part of the original sample of 500. Those in the first two rooms allow further consideration of Science Art as its own category and demonstrate the influence of the content of the art, the artist's style, and the medium used. Science Art images can be effective teaching aids, showing nature as it was and as it now is. Viewing images through the lens of science provides a sense of discovery, a sense of excitement, and a pleasure that increases when a scientific perspective—even one that involves some speculation about the content of the work or the intentions of the artist—blends with an appreciation of the medium and style of the work.

The images in the third room show the importance of captions—explaining why the combination of a work of art and its caption makes an effective tool for conveying truths about nature and for offering different views of environmental issues. Together, a work of art and its caption allow for the correction of errors—in the science portrayed, in the rendering of that science, or in the viewer's perception of both. In the fourth room we explore whether these images can be appreciated when we come upon them unexpectedly in public spaces or online.

The four paintings in the final room of the Upper Gallery summarize the value of Science Art. The first, *Young Lady in 1866* by Édouard Manet, provides an opportunity to view a narrative of culture that is linked to nature through the inclusion of a captive parrot. The darkened room containing the woman and her pet bird feels so quiet and removed from the bird's vibrant native habitat that it provides an emotional impression of its disconnection from nature. The picture affords a collateral opportunity to compare the terms "impressionism" and "Science Art." When the former was coined, some impressionists, like Manet himself, never applied it to their work. The second painting, *Ivory-billed Woodpecker* by George Miksch Sutton, as it appeared on the cover of *Science,* provides an opportunity to discuss the popularity of birds, the prominence of narratives of nature that address the human footprint on the planet, and the role of iconic animals, like the Ivorybill, that capture our collective imagination. The third painting, *Blackcocks in Springtime,* by David Klöcker Ehrenstrahl, which portrays a lek (mating assemblage) at a woodland edge, provides an opportunity to note again why humans are drawn to nature and why that attraction is so important. It also provides comparison with the last painting in this room, created 300 years later—a contemporary scene of a lek by Vadim Gorbatov. We do not know how Gorbatov painted an action taking place in the blink of an eye, but photography, and its ability to stop action, has helped many later artists find effective ways to portray dramatic events with fidelity. Contemporary viewers have themselves become accustomed, by experience with stop-action wildlife photography, to appreciate and understand paintings of this kind.

The two appendixes are intended to augment our gallery. The timeline in the first appendix broadens the context in which to consider the images, and the checklist in the second appendix conveys supplemental information on producing, exhibiting, and using Science Art in teaching about the environment. It also includes suggestions to facilitate finding artists and examples of their work.

Researchers have sought out birds to demonstrate connections among science, culture, and nature, in part because birds so sensitively reflect changes in environmental and cultural conditions and in part because birds have such broad appeal. Artists have sought out birds for similar reasons, though perhaps in reverse order. Both researchers and artists have been affected by the expanding public interest in avifauna. Birds are eye-catching and interesting, and watching them has become an avocation for hundreds of millions of people worldwide, including some 60 million in the United States alone. This has led the research to flourish: more than 3,600 bird-related reports were published in English in 2001.[3] And it has led the art to flourish: thousands of professional artists feature birds in their work—as reflected, for example, by the annual Birds in Art exhibition of the Leigh Yawkey Woodson Art Museum, now in its thirty-third year, which in 2007 included about 100 pieces chosen from nearly 1,000 entries (up from 589 entries four years earlier). Just as more artists are sensitive to the science underlying their narratives of nature, more viewers are now looking at art differently. Decoding the science is key, and this book demonstrates why.

Gallery Guide

LOWER GALLERY

PLATE 1. *Lower Gallery Map*, 2007, by Darryl Wheye (b. 1949). Computer graphic.

ROOM 1. Birds as Icons

PLATE 2. *Owl in Chauvet Cave*, Vallon–Pont d'Arc, France, c. 30,000 BCE. Finger tracing, height c. 119 in. (c. 3 m).

Of the thousands of Paleolithic cave wall images found so far, fewer than a dozen depict birds, but four of those depict owls. This image, produced 30,000 years ago, is the oldest. Even such simple images can convey clues about a bird's biology or ecological relationships.

PLATE 3. *Modified Owl in Chauvet Cave* [Eagle Owl], Vallon–Pont d'Arc, France, 1997/2007, by Darryl Wheye. Ink and colored charcoal drawing digitally placed over photograph, 12 × 14 in. (30.5 × 35. 6 cm).

The graphic emphasizes features of Plate 2.

PLATE 4. *Statue of King Chefren* (side view), c. 2500 BCE. Diorite, 37 × 22 × 66 in. (96 × 57 × 168 cm).

The falcon as an ancient Egyptian icon of insight and speed is included in hieroglyphics and works of art, such as those featuring sovereigns. Bird-sovereign associations occur elsewhere—for example, in the pre-Columbian image portraying Cinteotl, the Aztec Lord of Maize, who is accompanied by a parrot in the 800-year-old *Codex Laud* (MS. Laud Misc. 678 f. 9, 10).

PLATE 5. *Statue of King Chefren* (front view).

This is the view of the statue of King Chefren that was seen by the public. (See also Plate 4.)

PLATE 6. *The God Horus Protecting King Nectanebo II*, 360–343 BCE. Greywacke, 28 × 8 × 18 in. (71.1 × 20.3 × 45.7 cm).

In contrast to the understated falcon-sovereign association in Plate 4, in the association here, the bird overshadows the king, Nectanebo II. A biologically sound explanation can be made for this reversal.

PLATE 7. *Saint Gregory with Scribes* (detail), late 800s (Carolingian). Entire ivory relief, 8 × 4 8/9 in. (20.5 × 12.5 cm).

The dove murmuring as the saint writes is reminiscent of the falcon attending the Egyptian sovereigns Chefren and Nectanebo (Plates 4, 5, and 6) and the parrot attending Cinteotl. In Christian images a dove may also represent the soul; in secular images it may convey peace (or war), and good luck (or bad).

PLATE 8. *Farmer's Wife and the Raven*, 1782, by George Stubbs (1724–1806). Enamel on Wedgwood Biscuit Earthenware, oval 27 1/2 × 37 in. (69.9 × 94 cm).

Superstition and casting blame on ravens are the subjects of this painting and John Gay's late-eighteenth-century fable about a fallen horse and broken eggs. Raven lore is widespread. The birds are included in the creation myths of the Haida and those of various other cultures, in legends of ancient civilizations— in ancient Greece the bird so infuriated Apollo that he enlarged its bill and turned its feathers black—and in modern fables and tales. Often, as here, there is a biological basis for the behavior of the bird.

PLATE 9. *Winter Fields*, 1942, by Andrew Wyeth (b. 1917). Tempera on canvas, 17 1/4 × 41 in. (43.8 × 104.1 cm).

    A dead blackbird serves as a tribute to the fallen in war: Wyeth painted this bird lying on a senescent field in 1942 as World War II was expanding.

ROOM 2. Birds as Resources for Human Use

PLATE 10. *Netted Ostriches on a Libyan Outcrop,* Matendusc, near Murzuch (Fezzan), Libya, c. 8000? BCE. Rock engraving.

    This pictograph documents the hunting of ostriches 10,000 years ago in what is now Libya. These flightless, cloven-footed birds were also hunted in ancient Egypt and imported to ancient Greece and Rome, where they were displayed, ridden, used in circus games, fed to dogs in amphitheaters, and domesticated for food, feathers, skins, eggshells, and foot souvenirs.

PLATE 11. *Modified Netted Ostriches on a Libyan Outcrop,* Matendusc, 2005/2007, by Darryl Wheye. Pencil drawings digitally placed over photograph.

    The graphic emphasizes features of Plate 10.

PLATE 12. *Pelicans from a Wall Painting in the Tomb of Horemheb* (no. 78) [untitled wall painting] (detail), Thebes, Egypt, early 1400s BCE. Gouache on plaster over mud-straw foundation, 19 1/2 × 8 1/2 in (49.5 × 21.6 cm).

    An ancient Egyptian wall painting records the use of "domesticated" pelicans 3,300 years ago, during the reign of Tuthmosis IV. It raises questions about the practicalities of egg production in this species and the extent of domestication of other "wild" birds.

PLATES 13 and 13a. *Purse Lid from the Sutton Hoo Ship Burial* and detail, c. 600–650. Bars and panels of garnet and millefiori glass with filigree bindings, height of entire lid 41 in. (18.8 cm).

    A jeweled purse lid interred at the Sutton Hoo burial site on the English coast about 1,300 years ago shows a raptor and its duck prey, a motif that reappears in European art. Was this a symbol of authority, of the owner's dedication to falconry, or both?

PLATE 14. *Modified Detail of Purse Lid from the Sutton Hoo Ship Burial,* 1990/2007, by Darryl Wheye. Colored charcoal drawing, 11 × 14 in. (27.9 × 35.6 cm).

    The drawing emphasizes features of Plate 13a.

PLATE 15. *The Discreet Messenger,* undated, by François Boucher (1703–1770). Oil on canvas, 23 3/5 × 19 9/10 in. (60 × 50.5 cm).

    This painting by a French rococo artist brings attention to the long history of the pigeon post. The small ring on a pigeon's leg that once held handwritten messages can now hold microimages.

PLATE 16. *Happy Thanksgiving,* 2000, by Sally M. Berner (b. 1945). Oil on canvas, 24 × 30 in. (61 × 76.2 cm).

    This contemporary view of Thanksgiving turkeys on their way to market is a statement about the handling of market-bound birds in the multibillion-dollar poultry industry. It is also a reminder about the livelihood of those associated with poultry production, transportation, processing, and sales, which have come into focus through the threats associated with the spread of bird flu.

ROOM 3. Birds as Teaching Tools

PLATE 17. *Auk in Cosquer Cave,* Alpes Maritimes, France, c. 17,000–16,000 BCE. Charcoal on rock wall.

    A Paleolithic bird image on a cave wall serves as a contemporary indicator of global climate change. This cave art, now submerged beneath the Mediterranean, indicates that Great Auks were known to the artist 19,000 years ago, and places them far south of their modern range.

PLATE 18. *Modified Detail of Auk in Cosquer Cave,* Alpes Maritimes, France, 2005/2007, by Darryl Wheye. Ink, watercolor, and pencil drawings digitally placed over photograph.

    The graphic emphasizes features of Plate 17.

PLATE 19. *Animals Sketched from Nature (Xiesheng Zhenqin Tu)*, undated, by Huang Quan (c. 903–965). Handscroll, colors on silk, 16 1/5 × 27 1/3 in. (41.5 × 70 cm).

The artist used these studies, sketched directly from nature, to teach his son how to draw. They illustrate differences in the development of realism in Eastern and Western bird art. In both culture areas, artists used the work of other artists as source material. Compare the resources that Huang provided his followers with those provided by the monk who produced Plate 20.

PLATE 20. *Page from a Monk's Drawing Book* (pl. 11b), England, 1300s. Watercolor on vellum, 9.8 × 7.4 in. (24.9 × 18.8 cm).

Bird images modeled on art rather than nature may include, and even elaborate on, errors in the sources being copied. Compared with the birds in Huang's studies, which were drawn from nature and can be readily identified (see Plate 19), the species drawn in this monk's instructional guide often lack unique features, which makes them a challenge to name.

PLATE 21. *The Art of Flight, Daedalus (Il Volo di Dedalo),* from the Bell Tower (Campanile), Florence, Italy, c. 1334–1348, by Andrea Pisano (1270–1349). Marble relief.

The idea of human flight was often portrayed as the product of human arrogance: Icarus flew so high that the sun melted the wax binding his feathered wings, and he perished in the sea. Pisano's portrayal clearly shows the feathers Daedalus wore, calling attention today to the role of hind wings in *Archaeopteryx* and other transitional forms in the evolution of bird flight.

PLATE 22. *Portrait of Balthazar Sage,* 1777, by Colson (Jean-François Gilles, 1733–1803). Oil on canvas, 39 2/5 × 31 9/10 in. (100 × 80.9 cm).

Reliance on the canary in the coal mine is not fiction: birds serve as indicators of dangerous conditions, like the buildup of carbon dioxide or other toxic gases, or, by analogy, the presence of dangerous pesticides in the food chain. In 1777, Colson recorded a pharmacologist using passerines to demonstrate that the gas held in a flask was lethal.

PLATE 23. *Ahead of the Storm,* 1989, by Lee Stroncek (b. 1951). Acrylic on canvas, 22 × 29 in. (55.9 × 73.7 cm).

Migrating birds can indicate changes in the weather, and changes in their patterns of migration can indicate changes in climate. In birds, high metabolism, high mobility, and seasonal resource shortages led to the evolution of migratory behavior that synchronizes travel between breeding and wintering regions. Stroncek's painting of moonlit flight is a reminder that fall migrants must leave their breeding grounds early enough to minimize the risk of early storms but late enough for their wintering grounds to be well stocked for their arrival.

ROOM 4. Birds as a Means of Understanding Biology

PLATE 24. *Owls in Les Trois Frères Cave,* Ariège, France, c. 30,000–17,000 BCE. Stone carving, diameter 34 1/4 in. (87 cm).

The Paleolithic owl family etched on a cave wall could symbolize birth, rebirth, or the origin of life, like the goose nest found in King Tuthankhamen's tomb, or it could suggest a valued resource, like the lemmings, whose populations expand cyclically along with the owls who prey on them. Or perhaps the owls represent parental care, which in these monogamous birds lasts at least into each year's fall.

PLATE 25. *Ten Drachma Silver Coin,* Arkagas, in ancient Sicily, c. 412–411 BCE. The artist is unknown, but the letters "Poly" are evident on some coins. Engraved silver coin, diameter 1 1/2 in. (3.8 cm).

The eagles preying on a rabbit stamped onto a Greek coin could represent power or wisdom, like the falcon of ancient Egypt, the owl of Athena, and the eagle of Babylon, Imperial Rome, and the American West. The rabbit could serve as a reminder to take heed within range of danger, or it could represent—as it does today—the idea of fecundity.

PLATE 26. *Modified Detail of Ten Drachma Silver Coin,* Arkagas, Sicily, 1990/2007, by Darryl Wheye. Ink and colored charcoal drawing, 11 × 14 in. (27.9 × 35.6 cm).

The drawing emphasizes features of Plate 25.

PLATE 27. *Barnacle Geese,* in the *Harley Manuscript,* England, c. 1230–1240. Vellum, 12 × 9 1/5 in. (30 1/2 × 23.3 cm).

Unlike religious images of a pelican tearing open its breast (an allegory for self-sacrifice), scenes of tree-hanging Barnacle Geese seem to be associated with the mystery that surrounded their reproductive biology, a mystery that came into play during Lent. In some accounts the birds came from trees; in others, from barnacles clinging to floating timber or wooden vessels. (See also Plate 28.)

PLATE 28. *Goose-necked Barnacle* [*Lepas* sp.], after a drawing published by Ulisse Aldrovandi (c. 1639), and *Goose-necked Barnacle Cutaway,* after a drawing by Lois Rein, c. 1986, both by Darryl Wheye, 1996/2007. Pencil drawing, 5 × 6 in. (12.5 × 15.3 cm), and ink and colored charcoal drawing, 11 × 14 in. (27.9 × 35.6 cm), digitally combined.

The cutaway drawing exposes the feather-like feet of the Goose-necked Barnacle that were associated with the lore of the Barnacle Goose, and the sketch of the barnacle published by Ulisse Aldrovandi helps explain the origin of the lore.

PLATE 29. *The Parliament of Birds,* one of a pair of undated paintings, by Carl Wilhelm de Hamilton (1668–1754).

De Hamilton's painting is based on Geoffrey Chaucer's poem (1380) about mate selection, but it also serves as a virtual field guide, providing a record of local bird species and locally held exotic bird species known to the artist.

PLATE 30. *The Enchanted Domain I (Le Domaine Enchanté I),* 1953, by René Magritte (1898–1967). Oil on canvas, 26 3/4 × 53 3/5 in. (68 × 136.2 cm).

The painting provides a visual take on the coevolutionary relationship between seedeaters and the seed-producing plants upon which they depend. Viewers can extend this kind of relationship to hummingbirds and flowers, whose long bills fit into the nectar-producing flowers they pollinate, and oxpeckers and the tick-bearing mammals whose thick hides protect them from the birds' sharp, clinging claws.

ROOM 5. **Birds as a Means of Promoting Conservation**

PLATE 31. *The Shaft in Lascaux Cave* (detail), Montignac sur Vezere, France, c. 15,000–10,000 BCE. Pigment on rock wall.

Lascaux's two bird images and its only human image are found in the deepest portion of the cave. Certain identification of the species is impossible, but an egret and an eagle are possible contenders in addition to the often-cited raven.

PLATE 32. *Modified Detail of the Shaft in Lascaux Cave,* 1997/2007, by Darryl Wheye. Ink and colored charcoal drawing and watercolor painting digitally placed over photograph.

The graphic emphasizes features of Plate 31.

PLATE 33. *Coffin for an Ibis,* Egypt, possibly from Tuna el Gebel, 305–30 BCE. Gilded wood, silver, gold and rock crystal, 23 1/10 × 15 × 22 in. (58.7 × 38.2 × 55.8 cm).

The once-venerated Sacred Ibis is no longer found in Egypt; only extremely large numbers of mummified specimens in catacombs or museums remain. Why efforts to reintroduce the species have yet to succeed remains puzzling.

PLATE 34. *The Peasant and the Nest Robber,* 1568, by Pieter Brueghel the Elder (c. 1525–1569). Oil on oak wood, 23 1/10 × 26 1/3 in. (59.3 cm × 68.3 cm).

In much of the world, nest robbing is now illegal, but Brueghel's portrayal of the Flemish proverb "The one who knows the nest's location, / Can say that he has known. / The one who steals the nest, however, / Has it for his own" suggests that the popularity of harvesting wild bird eggs in the Netherlands during the sixteenth century could not be sustained even then. (The proverb is in Klein and Klein, *Peter Brueghel the Elder,* 168.)

PLATES 35 and 35a. *Untitled* [The banding of a heron] and detail, probably 1764, by an unknown artist.

Tagging dead birds collected in the wild in the 1600s to maintain a record of their origin led to banding live birds in the 1700s. This painting portrays the banding of a heron and could have catalyzed the spread of this new monitoring technique.

PLATE 36. *Mossy Branches—Spotted Owl*, 1989, by Robert Bateman (b. 1930). Acrylic on board, 16 × 20 in. (15.2 × 50.7 cm).

Bateman's painting contributed to the efforts to save the Spotted Owl in North America. The species became iconic in the 1980s during efforts to salvage remnants of old-growth forest from logging in the Pacific Northwest.

UPPER GALLERY

PLATE 37. *Upper Gallery Map*, 2007, by Darryl Wheye. Computer graphic.

ROOM 6. Science Art as Its Own Category

PLATE 38. *Sooty Tern and Birdman Petroglyphs on Boulder Panel 64, Mata Ngarau, Orongo, Easter Island*, 2005/2007, by Darryl Wheye. Pencil drawing, 8 × 11 1/2 in. (20.3 × 29.2 cm).

The birdman chimeras portrayed in this drawing were carved by members of an Easter Island cult that was active several hundred years ago. The Sooty Tern perched on the outcrop played an important role in the cult's annual spring election.

PLATE 39. *Birdman Petroglyphs on Boulder Panel 64, Mata Ngarau, Orongo, Easter Island*, 1982, photograph by Georgia Lee (b. 1926). Side of boulder shown in photograph, 87 × 54 in. (220 × 138 cm).

A comparison between the photograph and Plate 38 points out aspects of Science Art and shows the vulnerability of the unprotected carvings as erosion continues to wear away the boulders and destabilize the cliffs supporting them.

PLATE 40. *Honeybees, Honeyguides, and Honey Hunters*, 2004/2007, by Darryl Wheye. Pencil drawing, 25.5 × 8 in. (64.8 × 20.3 cm).

Honeybees, Greater Honeyguides (an African bird), and honey hunters of the Boran tribe share an interest in honey. The honeyguides and the hunters team up to raid the bees.

PLATE 41. *Twittering Machine (Zwitscher-Maschine)*, 1922, 151, by Paul Klee (1879–1940). Oil transfer drawing and watercolor on paper [mounted] on cardboard, 16 1/4 × 12 in. (41.3 × 30.5 cm).

Klee's painting seems to anticipate the invention of the sonogram.

PLATE 42. *Drawing Based on a Detail from Paul Klee's "Twittering Machine" with Sonogram*, 1995/2007, by Darryl Wheye. Sonogram elements added to pen and ink drawing using Photoshop, 8 1/2 × 11 in. (21.6 × 27.9 cm).

The drawing emphasizes features of Plate 41 that support the sonogram interpretation.

PLATE 43. *Model of Flying Machine Driven by the Force of Man by Leonardo da Vinci*, c. 1505 / 1988, constructed by James Wink, Tetra Associates, London, based on Ms.B.f.74v and others. Beech wood, iron, brass, hemp and coir rope, tarred marline, leather, and tallow. Wing span 433 in. (1,100 cm); length 126 in. (320 cm); depth 66 in. (167 cm); depth of wing stroke 193 in. (490 cm).

This model of one of Leonardo's flying machines is remarkable for both its design and its weight. Although it looks like an ultralight aircraft, it is ultraheavy.

PLATE 44. *An Experiment on a Bird in the Air Pump*, c. 1768, by Joseph Wright of Derby (1734–1797). Oil on canvas, 72 × 96 in. (182.9 × 243.8 cm).

Wright's painting records not only an experiment but also the social setting in which it was performed: it portrays after-supper science, which was a monthly draw for members of the Lunar Society.

PLATE 45. *Ascending and Descending*, 1960, by M. C. Escher (1898–1972). Lithograph, 14 × 11 1/4 in. (35.6 × 28.5 cm).

Escher's often-discussed lithograph demonstrates one advantage of collaborative efforts between artists and scientists.

PLATE 46. *Snake and Small Bird*, c. 1836, by Hokusai (Katsushika Hokusai, 1760–1849). Color on paper, 10 7/10 × 9 1/3 in. (25.6 × 23.6 cm).

Hokusai's painting illustrates the role of aesthetics and the power of visual metaphors in representing nature.

PLATE 47. *Muscles That Raise and Lower the Lower Mandible*, 1984, by Tony Angell (b. 1940). Ink on scratchboard, 3 1/2 × 3 1/2 in. (9 × 9 cm).

This drawing of bird skulls and muscles shows how two kinds of blackbirds raise and lower their mandible, serving as a classic example of the value of illustrations.

PLATE 48. *Male Black-backed Oriole with Captured Wintering Monarch Butterfly in Pine Woods above Mexico City,* 1984, by Tony Angell. Ink on scratchboard, 8 × 6 3/4 in. (20.3 × 17.2 cm).

This second image of blackbirds by Tony Angell shows some of the differences between pictures that are usually seen as illustrations and pictures that are labeled art. It also shows one of the traditional differences between what can be called generic art and Science Art: quantity of information.

PLATE 49. *Woman Observing Bird (Willie Was Different),* 1967, by Norman Rockwell (1894–1978). Oil on board, 16 × 8 1/8 in. (40.6 × 20.6 cm).

Rockwell's painting shows another traditional difference between narratives that are illustrations and narratives that are art: convention.

PLATE 50. *Brewer's Blackbird,* c. 1914, by Louis Agassiz Fuertes (1874–1927). Watercolor on paperboard, 11 × 15 in. (27.9 × 38.1 cm).

Fuertes' painting shows a third perceived difference between illustrations and art: venue.

ROOM 7. Content, Style, and Medium

PLATE 51. *The Raven,* 1995, by Henry Bismuth (b. 1961). Oil on canvas, 44 3/4 × 76 3/4 in. (113.7 × 195 cm).

The content of a picture can present the behavior of an individual. Here, Bismuth's painting shows the exertion expended by calling ravens.

PLATE 52. *Picnic,* 1997, by Paula G. Waterman (b. 1954). Oil on canvas, 22 × 28 in. (55.9 × 71.1 cm).

The content of a picture can also present intraspecies behavior. Here, Waterman's painting records the commensal relationship between these ravens and people, in which the birds benefit, but not at a cost to the people providing the food.

PLATE 53. *Without Warning,* 1998, by Carl Brenders (b. 1937). Gouache and watercolor on illustration board, 26 1/2 × 38 1/4 in. (67.3 × 97.2 cm).

The content of a picture can present interspecies behavior as well. Here, Brenders's painting gives us a bird's-eye view of a raven chasing a young eagle.

PLATE 54. *Evening Grosbeak,* c. 1948, by Roger Tory Peterson (1908–1996). Lithograph, 14 2/3 × 11 3/5 in. (37.2 × 29.5 cm).

The content of a picture can also present opportunities to discuss changes in community ecology and habitat, even when they are not depicted. Here, Peterson's painting portrays a species whose range in North America expanded along with the human practice of hanging up backyard bird feeders, and then changed again, presumably in response to the human-derived effects of global climate change.

PLATE 55. *Champion,* 1998, by Patricia Pépin (b. 1964). Oil on canvas, 12 × 24 in. (30.5 × 61 cm).

Finally, the content of a picture can present —or allude to—human-animal interactions and their outcomes. Here, Pépin's painting serves as a platform for considering the impact of the automobile on avian habitat and the truth of the saying "Nature bats last."

PLATE 56. *New Ark,* 1991, by Ray Harris Ching (b. 1939). Oil on composition board, 27 1/2 × 36 in. (69.9 × 91.4 cm).

An example of a painting done in a realistic style, Ching's *New Ark* portrays a contemporary ark whose passengers, for the most part, are asleep.

PLATE 57. *The Tyrants,* 1997, by Thomas Quinn (b. 1938). Watercolor on gessoed paper, 24 1/4 × 15 1/2 in. (61.6 × 39.4 cm).

An example of work done in a painterly style, Quinn's painting features a large owl and a little hummingbird, opposites that are, in a way, two of a kind.

PLATE 58. *Long-billed Curlew,* Palo Alto Baylands, California, 2004, by Tom Grey (b. 1941). Photograph taken using the Canon 300mm f/4 IS, with a Canon 1.4× extender attached, making a 420mm f/5.6 (1130 × 750 pixel cropped from a 3072 × 2048 original).

An example of a photograph as a form of art, Grey's *Long-billed Curlew* presents a clear view of this stocky, big-billed bird—mud included.

PLATE 59. *Long-billed Curlew,* 1998, by Chris Bacon (b. 1960). Watercolor on rag-board, 15 1/2 × 15 in. (39.4 × 38.1 cm).

An example of painting as a form of art, Bacon's *Long-billed Curlew* presents a significantly different take on this species compared with that in Grey's photograph (Plate 58).

PLATE 60. *Eskimo Curlew,* 1957, by Charley Harper (1922–2007). Serigraph, 18 1/2 × 13 in. (46.3 × 32.5 cm).

An example of minimal realism, Harper's *Eskimo Curlew* serigraph presents an unassuming composition that nonetheless tells us a lot about a curlew species.

ROOM 8. The Importance of Captions

PLATE 61. *Two Stories—Common Nighthawk,* 1994, by Carel Pieter Brest van Kempen (b. 1958). Acrylic on illustration board, 20 × 30 in. (50.8 × 76.2 cm).

Brest van Kempen's painting, besides providing an interesting point of view, presents the elements of a good Science Art caption and a good Science Art catalogue entry.

PLATE 62. *Northern Mockingbird [Mocking Bird] (Plate 21),* c. 1825, by John James Audubon (1785–1851). Watercolor.

This Audubon painting plus caption shows how captions can be used to rectify errors.

PLATE 63. *Brown Thrasher [Ferruginous Thrush] (Plate 116),* 1829? by John James Audubon. Watercolor, graphite, pastel, and black ink with scratching out and selective glazing on paper, laid on thin board.

In contrast to the preceding Audubon painting (Plate 62), this one carries with it neither the cost nor the benefit of including an error and helps to convey the value of captions.

ROOM 9. From Real Public Venues to Virtual Ones

PLATE 64. *Vultures and Crystals* [Oriental White-backed Vultures (*Gyps bengalensis*)], 2004/2007, by Darryl Wheye. Pencil, 12 × 30 in. (30.5 × 77 cm).

This portrayal of the massive die-off of Oriental White-backed Vultures in India in the 1990s was included in a short-term exhibit in Stanford University's Falconer Biology Library.

PLATE 65. *Avian Engineering* [Bushtits (*Psaltriparus minimus*)], 2005, by Darryl Wheye. Watercolor, 12 × 20 in. (30.5 × 50.8 cm).

This watercolor highlighting the structure of a bird's nest has been installed in a public space frequented by engineering students.

ROOM 10. Science Art, Birds, and Perceptions of Nature

PLATE 66. *Young Lady in 1866,* 1866, by Édouard Manet (1832–1883). Oil on canvas, 72 8/9 × 50 2/3 in (185.1 × 128.6 cm).

Manet's painting is a narrative of culture rather than nature. It was produced when city dwellers saw mostly horses, market animals, pigeons, and dogs. (See N. Hammond, *Modern Wildlife Painting,* 12.) The inclusion of an exotic parrot in a painting by a major artist, even when portrayed as an accoutrement, can affect the popularity of the species and increase risks to the wild population.

PLATE 67. *Ivory-billed Woodpecker,* c. 1935, by George Miksch Sutton (1898–1982). Watercolor, as used on full cover of *Science* (June 3, 2005). Cover size 10 1/2 × 8 1/4 in. (26.6 × 21 cm).

Sutton's painting appeared on the cover of *Science* in the June 2005 issue, which published the report of the bird's apparent rediscovery. The painting attests to the popularity of rare wild birds and of public interest in events that surprise scientists.

PLATE 68. *Blackcocks in Springtime (Orrspel),* 1675, by David Klöcker Ehrenstrahl (1629–1698). Oil on canvas, 108 × 105 in. (275 × 266 cm).

Ehrenstrahl's painting shows a lek: an assembly area where grouse engage in courtship activities. The male grouse, in hopes of winning mates, contest for territories that females prefer.

PLATE 69. *Attack, Out of the Mist,* 1995, by Vadim Gorbatov (b. 1940). Watercolor, 16 1/2 × 11 4/5 in. (41.9 × 31.5 cm).

Compared with Ehrenstrahl's lek (Plate 68), this one by Gorbatov shows how the addition of a predator heightens the tension and suggests that the perception of high-speed action benefits from stop-action photography.

Lower Gallery
Bird Art over the Millennia

The images of an owl etched into the soft clay of a cave wall 30,000 years ago, included in Egyptian hieroglyphics, woven between the words of a bestiary during the Middle Ages, painted on canvas in the eighteenth century, molded out of plastic in the twentieth, and saved in bytes on a laptop are all related to one another. Similarly, the images in each room in the Lower Gallery illustrate a single topic and are presented systematically, allowing consideration of possible patterns in the history of their production. In each room, too, the examples were produced in roughly the same time periods, beginning with one from prehistory, followed by one from an ancient civilization and one from the Middle Ages or Renaissance. Next come examples produced around the early Scientific Revolution (the late 1600s and early 1700s) and then during the buildup to the Information Age (beginning in the mid-1900s).[1] The first seventeen examples show how humanity has valued birds as icons, resources, and teaching tools; and the following eleven feature aspects of bird biology and bird conservation.

In many works of ancient art we refer to the birds portrayed using nonspecific terms like falcon or dove, rather than pinpointing the species. Sometimes, as with the Chauvet owl (see Plate 2), we discuss the species name to widen the discussion, but in most cases trying to identify a bird at the species level would introduce inaccuracies; after all, historical ranges change, often with climate. The bird in Plates 4 and 5, for example, is probably a member of the falcon family (Falconidae), of which eleven species can be found in Egypt today. The sculpture (c. 2500 BCE) might represent a Lesser Kestrel (*Falco naumanni*), or a Eurasian Kestrel (*F. tinnunculus*), or a Red-footed Falcon (*F. vespertinus*), or an Eleonora's Falcon (*F. eleonorae*), or a Sooty Falcon (*F. concolor*), or a Merlin (*F. columbarius*), or a Eurasian Hobby (*F. subbuteo*), or a Lanner Falcon (*F. biarmicus*), or a Saker Falcon (*F. cherrug*), or a Peregrine Falcon (*F. peregrinus*), or even a Barbary Falcon (*F. pelegrinoides*). We can probably narrow the choice down to eight species with some confidence because, with the exception of the Eleonora's, Sooty, and Barbary Falcons, each species has been identified from mummified specimens. It should be noted, however, that most of the thousands of falcon mummies are from a much later date—2,500 years later—when falcon ranges could have been significantly different from those during the reign of Chefren. On the other hand, at the time of Chefren, the ancient Egyptians might have clumped eagles, hawks, and kites together with falcons. Identifying the Chefren falcon, then, requires more knowledge about the cultural role of falcons and the historical range of individual species than is currently available.[2]

Room 1 features six images that show a few of the birds that became icons. The range of iconic birds is wide; they represent everything from peace and power (dove and raptor) to foolishness and sagacity (dodo and owl). Surprisingly often, the qualities they are honored for have a biological basis.

Room 2 features five images that display a few of the birds we humans have valued—and often still value—as resources. In addition to food, birds have represented many forms of wealth and status, supplying everything from ornate apparel and hunting aids

(Hawaiian feather robes and falconry), to a means of communication (homing pigeons), protection against the cold (down lining), and companionship.

Room 3 features six images that show some ways in which birds have served as teaching tools. Birds have been models for replicating flight for ancients and moderns alike (from Daedalus and Leonardo da Vinci to modern airframe designers who study the dynamic soaring of vultures), for understanding evolution (from Darwin's finches to studies of systematics), for monitoring the environment (from the miner's canary to keystone species, used in determining ecosystem quality), and for monitoring disease vectors (from the West Nile virus to avian influenza).

Room 4 features six images that show ways in which artists have recorded aspects of avian biology. The selection barely scratches the surface, for avian biology is such a rich and productive discipline that it would be impossible to cover even in a sample ten times the size.

Room 5 features five images that reflect the recognition of limits, from drawings on the walls of Lascaux to paintings of birds under legal protection.

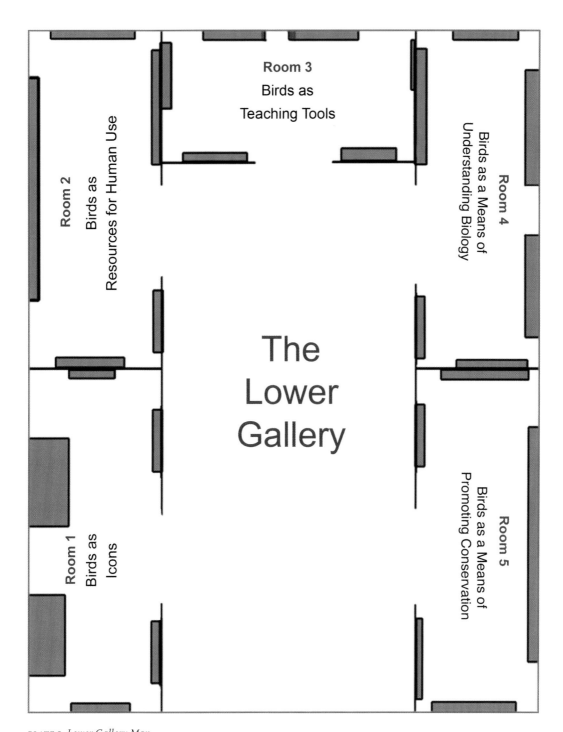

PLATE 1. *Lower Gallery Map*
This gallery contains 26 original images that, if found in a single collection, might be housed not by culture or by era but by topic and separated into rooms as shown here.

ROOM I

Birds as Icons

The history of iconic birds serving as intermediaries to the gods or as ostensible models of human behavior is long and little known, but clues are noted in the captions.

The history of iconic bird masks and bird chimeras, both of which were sometimes venerated, is almost as long. A Paleolithic artist included what is usually interpreted as a bird mask in Lascaux, 12,000 years before pictographic writing was devised (see Plate 31), and much more recently, within the past few hundred years, an artist in New Guinea added a carved merganser (ducklike waterfowl) to his mask, and an artist in Burkina Faso added a large, carved decurved bill to his. The inclusion of bird parts in these three masks has often been interpreted to mean that the wearer relied on intervening guidance.

But iconographic chimeras, like sphinxes, griffins, winged horses (Pegasus), angels, and cherubs—the whole array of imagined beasts with incongruous parts—have been used to convey an amalgam of traits, and when the parts include bird wings and heads, the chimeras also suggest speed and vision. They are, however, far from biological reality. So are the fire-emitting, winged dragons of Europe, even if these part-bird, part-reptile, warm-blooded(?) chimeras might portend a developing understanding that birds are feathered dinosaurs.

Just as certain live birds, like owls, eagles, falcons, ravens, doves, and even herons, have been used as metaphors for celestial intervention or for avian and human behavior, certain dead ones, like ducks and doves, have been used to signify wealth, prey, or protest. In some cases, as with the heron in ancient Egypt, which was thought to have sprung from the Nile and was associated with the origin of life, the role as a metaphor seems to be restricted to a particular place and set of conditions. In others, as with the eagle, the metaphors seem to be variable and widespread. In mythological and religious images, eagles often link humanity to events beyond human comprehension or control. In Roman myths they typically represent Zeus (plucking the liver from Prometheus or abducting Ganymede, son of the founder of Troy). In Christian images, eagles often represent John the Baptist; in Native American images they often represent the Thunderbird. Eagles also have their secular uses. They have appeared on money (see Plate 25) and flags for millennia and been linked to political individuals or political events. Leonardo, for example, used one to represent Francis I, who had ambitions to become the next Holy Roman Emperor.

Few are surprised that eagles, large and dramatic as they are, have had a long history as icons. Yet even songbirds have been icons. The Goldfinch (*Carduelis carduelis*), a popular cage bird in Europe and probably the most familiar bird in Christian paintings, is associated with the soul and with resurrection. During the plague outbreak in Europe (1347–1351) it was linked to healing and renewal, an iconic meaning that held for more than a century after the outbreak dissipated. The bird is still featured in forty-one of fifty-one extant Venetian paintings of Mary and Jesus produced between 1475–1525, but by the mid-1700s most viewers would not have known the history of the icon and would not have made the healing and renewal interpretation.

PLATE 2. *Owl in Chauvet Cave*, Vallon–Pont d'Arc, France, c. 30,000 BCE

PLATE 3. *Modified Owl in Chauvet Cave* [Eagle Owl], Vallon–Pont d'Arc, France, 1997/2007, by Darryl Wheye

PLATES 2 AND 3   **An Owl as Cave Art**

Art

In 1994 news of the discovery of Chauvet Cave in France and its walls full of Paleolithic art spread rapidly, and photographs of the woolly rhinoceroses, lions, bears and other animals, including this owl, were almost immediately available to accompany the early reports. Photographs of the owl were not as widely circulated as were those of the grotto's megafauna, but Paleolithic bird images were not as widely produced as the megafauna were, either. Years earlier, André Leroi-Gourhan (1911–1986), former director of the Museé de l'Homme in Paris, had surveyed seventy-two caves in France and neighboring countries and listed over 2,000 animal images on the walls. Among them, horse images outnumbered the rest, with 610; bison followed with 510; and mammoths came in a distant third with 205. Fish accounted for a paltry 8, and birds (or their heads) for only 6 (2 in Lascaux, 4 in Les Trois Frères).[3]

The official cave Web site describes the Chauvet owl image as follows: "This finger tracing represents an owl. The position of the wings shows that its head is turned 180 degrees relative to its posterior face. The anatomical characteristics of the animal permit its attribution to *Moyen Duc* [Long-eared Owl] (*Asio otus*). This drawing was realized on the soft outer layer of the cave wall. In the background we see traces that show the wall surface was scraped before the drawing was made."[4]

Science

Evaluating avian taxonomy based on an ancient figure etched into the soft surface of a cave wall is a difficult business, and an alternative interpretation is available. It may be that the bird is facing forward and that the species might actually be an Eagle Owl (*Bubo bubo*). Given Paleolithic artists' fascination with large and imposing creatures, that conclusion strikes the authors as more plausible. To emphasize the resemblance between the cave image and the Eagle Owl, we have inserted a drawing of its head into the accompanying image (Plate 3).[5]

Other factors—the relative size of the birds, their presence in the cave, and the suite of other cave images—support the alternative hypothesis. Eagle Owls are the size of eagles and prey on the smaller Long-eared Owls. They nest in caves, whereas Long-ears typically seek abandoned crow nests, found mostly in trees or shrubs. They are also impressive predators: they have been known to take down roe deer, and such accomplishments have earned for members of their genus (*Bubo*) a reputation as "Tigers of the Air." In contrast, Long-ears are more commonly preyed upon by others in their tribe. Long-ears have even evolved a defensive posture: They lean forward while arching their wings—like a swimmer poised to dive from a racing block. Then, by lifting the trailing edge of their wings skyward, they frame their face and appear more formidable, the way "owl eyespots" on the wings of certain moths may fool predators into thinking they are too large to take.[6]

The message of a Paleolithic artist is unknowable, and the message of any work of art can change with time. The modified image (showing details of the head) and this discussion do not resolve the identity of the species, but they do raise the issue and provide today's viewers with additional information, allowing them to make the comparisons and consider alternatives for themselves.[7]

PLATE 4 *(above). Statue of King Chefren* (side view), c. 2500 BCE

PLATE 5 *(top right). Statue of King Chefren* (front view)

PLATE 6 *(right). The God Horus Protecting King Nectanebo II*, 360–343 BCE

PLATES 4, 5, AND 6 Falcons and Egyptian Kings: Relative Size Matters

Art

In ancient Egypt, the king was a personification of the god Horus, the sky god, represented by a falcon. That falcon appears in both of these statues, which are about power: the falcon that conveys it and the viewer's perception of it. Chefren (Plates 4 and 5) built the second pyramid at Giza and the Sphinx. Nectanebo II (Plate 6), the last Egyptian pharaoh, ruled two millennia later as the civilization was winding down.[8]

The statue of Chefren radiates authority, even as he smiles above his fake ceremonial beard. Two lions (symbols of power and protection) are incorporated into the base of his throne along with a knotted papyrus representing Lower Egypt and a lily representing Upper Egypt (a hieroglyphic message of unity). The strength of Chefren's command is conveyed through his relationship with Horus, seen here as the life-size falcon perched behind his head, with its wings spread in a further symbol of protection. The falcon is strategically placed so that when viewed from the side, Chefren appears to be operating under its protective guidance. But when he is viewed from the front, the bird is hidden, so that approaching subjects will see Chefren as all-powerful. In contrast, in the statue of Nectanebo II, Horus dwarfs the king, and it is the falcon's double crown that represents authority over Upper and Lower Egypt.

Ancient Egyptian civilization depended on the moderate annual flooding of the Nile. A margin of 5 feet (1.5 meters) was the difference between the river failing to overrun the bank and floodwaters washing away the irrigation system. Naturally, the failure and success of the floods shaped regal influence. As regal influence shifted over time—in response to floods and other factors—so did the relative size of the falcon in commemorative artwork. Chefren's reign was prosperous as measured by the extensiveness of his major constructions and the wealth represented by the contents of private tombs. Nectanebo II, on the other hand, ruled at a time of collapse. The modest image seen here might suggest that he is handing over responsibility for Egypt's collective troubles to the gods.[9]

Science

The ebb and flow of Egyptian civilization was closely linked to the effect of extreme weather on the flow of the Nile and thereby to annual harvests and the resulting density of the populations of small migratory birds and other vertebrates upon which falcons based their diet. The pattern of occasional droughts and catastrophically high floods allowed the natural history of falcons to figure prominently in the lore of this agricultural society: when the falcons were absent, so too was prosperity.[10]

PLATE 7. *Saint Gregory with Scribes* (detail), late 800s

PLATE 7 **The Dove as Messenger**

Art

Sixteen popes have been called Gregory. The one seen here, also known as Saint Gregory or Gregory the Great (c. 590–604), was the first monk to become pope—and the one after whom the Gregorian Chant is named. A prolific writer, he is often shown with a dove, as he is in this carved ivory relief found on the cover of a ninth-century German manuscript. He once commented, "What Scripture is to the educated, images are to the ignorant."

The scene shown here calls to mind a story that his associate Peter the Deacon recorded in *Vita* (xxviii). The tale goes roughly like this: Just before dictating his discourse on Ezechiel, Pope Gregory drew a curtain between his secretary and himself. The dictation followed, but was halting in a way that concerned the secretary. Finally, the repeated pauses unsettled the secretary so much that he peeped through an opening in the partition. To his surprise, the pope appeared to be taking dictation himself—and it was coming from a dove. The idea struck a chord in those who heard the story, perhaps because of the biblical account of the dove and the ark and the bird's role as scout.

Science

The dove perched on the shoulder of Gregory the Great as he writes is reminiscent of the falcon perched on the shoulders of Chefren as he ruled (see Plate 4). But assigning a biological basis for the iconic dove requires more speculation than assigning one for the falcon does. Perhaps the selection of a dove could relate to the quiet quality of its voice; the murmuring of the dove surely is more suggestive of speech than, say, the trill or warble of a songbird or the cry of a raptor. In the teachings of the medieval Christian church, a quiet voice implied compliance. Hugh of Fouilloy in his *Book on Birds*, written between 1132 and 1152, says, "Instead of a song the dove uses a sigh, because by wailing it laments its willful acts." Hugh also credits the dove with joining flocks, forgoing predation and stealing, and raising "twin chicks." Medieval science in Hugh's day swept up some interesting direct observations, but it also included a great deal of mythology and superstition.[11]

PLATE 8. *Farmer's Wife and the Raven*, 1782, by George Stubbs

PLATE 8 **Ravens as Omens:
Some Biological Underpinnings**

Art

An old superstition, one apparently going back for millennia, warned that it was bad luck to hear a raven call from the left. The Roman dramatist Plautus (c. 254–184 BCE) made note of it in *Aulularia* (act iv, scene 3): "It means something—that raven cawing on my left just now!"[12]

Wariness after hearing a raven call from the left is the subject of this painting. It takes its title from a work by John Gay, who made the "old wives' tale" into an eighteenth-century fable, *The Farmer's Wife and the Raven*. In Gay's fable, the bird explains that disaster is neither caused by nor foretold by ill omens. The story goes as follows: While taking eggs to the market, a farmer's wife hears a raven call from the "unlucky side of the road" and concludes that trouble lies ahead. Her fears materialize when her horse stumbles and the eggs fall out of her basket and break. She blames the raven for the loss of the eggs:

> "That raven on yon left-hand oak
> (Curse on his ill-betiding croak)
> Bodes me no good."

The unperturbed bird asks:

> "But why on me those curses thrown?
> Goody, the fault was all your own."

When she points out that the bird spooked her horse, the bird faults her for choosing one ill suited to the task.[13]

Science

What makes a raven croak? Many things can induce vocalization in birds, and members of the crow family—like the raven—can be garrulous. In this case, however, the sight of a sizable cache of eggs could well have stimulated the bird to send out a vocal alert to fellow ravens. These birds share foraging information when they discover a good amount of food, like a carcass. By scouting separately, members of a group can improve the chances of spotting such sporadic finds; pooling information is an effective strategy, and birds would be likely to remain within earshot.[14]

For birds that roost colonially and seek large food objects—the large vultures of Africa's Great Rift Valley present another example—sharing information about a big find is a way of making a large resource available to the group. The biologist Colin Pennycuick once studied White-backed and Ruppell's Griffon Vultures (*Gyps africanus* and *G. rueppellii*) using a small, motorized sailplane. Sometimes while taking one of the thermals over the Serengeti Plain, he would notice a vulture at about his visual limit in one direction and another in the other direction. When he landed to eat his sandwich, he was surprised at the crowd of vultures that accumulated around him one by one. They must have examined the pilot in some dismay, wondering where the carcass was. The first finder is usually not a biologist, of course, but a vulture, and the find is usually large enough to share.

The British painter and engraver George Stubbs, who produced this image, was the son of a currier and leather salesman. Virtually self-taught, Stubbs specialized in animal images, particularly horses. He painted the scene shown here on earthenware produced for him by a Lunar Society member, Josiah Wedgwood. The Lunar Society was founded in England (see Plate 44). Members met informally each month to discuss science and watch experiments: "They caught at discoveries with delight, sure that every find could help them to crack the elusive codes of nature." Wedgwood, for example, is usually described as a potter, but he also devised a high temperature thermometer for use in his kiln that won him election to the Royal Society.[15]

PLATE 9. *Winter Fields*, 1942, by Andrew Wyeth

PLATE 9 **A Dead Crow as an Icon**

Art

One winter day Andrew Wyeth found a frozen crow while walking in the fields in Chadds Ford, Pennsylvania. In 1942 he painted that crow, and in 1943, while the war in Europe was raging, the painting was exhibited in the Museum of Modern Art in New York. Europe's crow is called the Carrion Crow (*Corvus corone*), and the death of the bird that cleans battlefields may not have been lost on wartime viewers of the painting. In fact, the strife of the 1940s appears to have prompted Wyeth to produce numerous sober images; many of his works from this period include dead birds, shriveled vegetation, and buildings abandoned or left in ruin, and one is a portrait of a Turkey Vulture (*Cathartes aura*) in flight. Wyeth sought to eliminate technical mannerisms that could stand between his expression and the viewer, and precise, detailed scenes like *Winter Fields* tend to evoke loneliness and remembrance.[16]

Science

The biological messages in this image are seemingly straightforward: Life is temporary, its requirements are exact, and for a lucky few, senescence will precede death. Viewers today would probably not be surprised if this painting were a commentary by Wyeth on World War II. Now, however, the crow symbolizes a quite different war: urban and suburban crows are the most prominent and visible victims of West Nile virus, which has become epidemic in some populations of U.S. birds. Because the virus is also a threat to humans and domestic animals, it is a serious public health concern. Perhaps the Wyeth painting will someday see service in a public information campaign warning of the spread of the West Nile virus or another bird-related disease. The new field of movement ecology discovers more and more about how physiology, evolution, behavior, and environmental forces determine where birds, and the viruses they sometimes harbor, travel.[17]

ROOM 2

Birds as Resources for Human Use

Many species of birds have been hunted, plucked, domesticated, kept as pets, trained to seize prey, or bred to fight others of their kind. Artists have left a record of advances in our methods of hunting, training, and utilizing them. Artists have also left a record of wild birds that provide free services, such as consuming insect pests upturned by farm machinery, pollinating flowers while feeding on nectar, dispersing seeds that they swallowed whole (see Plate 30), and serving as a cleanup crew by scavenging carrion. The narratives of birds providing these free services make up only a small portion of the pictorial record, however. Louis Agassiz Fuertes' painting of a wild blackbird holding a crop pest in its bill (see Plate 50) tells an important story, but according to the artist's daughter, Mary Fuertes Boynton, Fuertes realized the subject would have been difficult to market to a private collector. He was an artist trying to make ends meet during World War I, so if the U.S. government had not commissioned it, Fuertes would probably have been obliged to select a different subject.[18]

The extent of our age-old reliance on birds is easy to underestimate, even with an extensive pictorial record. The birds we commonly think of as domesticated (chickens, ducks, geese, swans, and pigeons) have been raised for thousands of years. Even quail have been raised for at least 1,500. Some of these utility birds regularly appear in the pictorial record; the appearance of others there, like the pelican in ancient Egypt (see Plate 12), comes as a surprise. The line between "wild but often used" and "domesticated" is blurrier than might be expected. To make the case for domestication, we would need evidence that human influence had brought about genetic changes that created forms or behaviors more useful to people than were found in the wild bird. Similarly, images of such enterprising activities as collecting down for padding might be fairly common but are not widely known, whereas various portraits of falconers and their birds (see the frontispiece) and scenes of fighting cocks—for instance, William Hogarth's *Pit Ticket: The Cockpit,* published in 1759, and François Boucher's message-bearing homing pigeon (see Plate 15)—are probably known to many.

According to the pictorial record, songbirds were caged and kept in ancient Egypt, Greece, and Rome, and birds, usually local passerines, were tethered or held in captivity. Artists' portrayals of flashy exotic (nonnative) species like parrots surely increased their popularity and led to changes in their wild populations (see Plate 66). Humans have introduced at least forty-eight parrot species to new areas, naturalizing about twenty, and left many native populations under threat due to overhunting for the pet trade and habitat degradation.

Parrots were mentioned by Aristotle. So were peacocks. Peacocks were even mentioned as objects of trade in the Bible: "For Solomon's navy ... once in three years went across the sea to Tharsis, and brought from thence gold, and silver, and elephants' teeth, and apes, and peacocks" (2 Chronicles 9:21). They were dined on in Imperial Rome—Varro wrote that Quintus Hortensius was the first to serve the meat at the banquet table, and Cicero described consuming it. Their feathers were plucked for garlands and fly swatters. By the 1300s, the birds were found in France, Germany, and England, where they were held captive in gardens and parks. Their appearance, and that of other exotic birds, in paintings produced during the Age of Exploration often coincides with voyages of various seafaring countries whose ships returned home with live specimens for private menageries and zoos and with less animated specimens for collectors.

Hunted birds are well represented in the pictorial record. Before firearms were invented, birds were grabbed by hand, knocked down with sticks, or caught in birdlime (a sticky substance applied to branches, sometimes placed near tethered owls, which the hunters knew other birds would mob). Birds were also hunted with spears, arrows, and slingshots, trapped using nets, and seized by trained birds of prey. Like falconry, firearms extended the range of the hunter and eliminated the hunter's need to kill the bird personally. In the 1600s and 1700s, during the Dutch golden age of still life painting, many artists included birds in their work; those shown as trophies usually look as pristine as they had in life.

PLATE 10. *Netted Ostriches on a Libyan Outcrop*, Matendusc, near Murzuch (Fezzan), c. 8000? BCE

PLATE 11. *Modified Netted Ostriches on a Libyan Outcrop*, Matendusc, 2005/2007, by Darryl Wheye

PLATES 10 AND 11 **Netting Ostriches in Neolithic Africa**

Art

The recorded history of ostrich use—in both the pictorial and the written archives—is long. The Kalahari Bushmen included them in their carvings and folklore. The Assyrians considered them holy, as did the ancient Egyptians, who used an ostrich feather to represent the goddess Ma'at. The Bible prohibited their consumption, but Persian kings ate them. So did the Roman emperor Heliogabalus (second century), who is said to have served 600 ostrich brains at a banquet. Greeks and Romans rode them. Herodotus reports that in Libya the Macae used their skins as shields. Diodorus Siculus, Xenophon, Herodian, Claudian, Lucian, Synesius, Catullus, Oppian, and Pliny also wrote about them. Ostriches were hunted for display in the circus. The Roman emperor Commodus (161–192) is said to have slaughtered them for public amusement "with crescent-shaped arrow-heads that neatly beheaded the creatures, while their bodies went running on." And they were hunted for their plumes, which were included in the wardrobes of women of fashion, inserted into the helmets of generals, and used to decorate the ceremonial headgear of officers and their horses.[19] Commercial ostrich ranching was started in South Africa more than a century ago and has moved to the American Southwest, where there are large-scale ranches managing thousands of birds. The early uses involved their decorative feathers; more recently, their hides have been valued—for shoes and other items. Now a significant experiment in raising the birds for meat is under way.

In the wild, these diurnal birds are loosely gregarious and travel widely. When pursued, they lope steadily, but they have a tendency to break into a sprint and zigzag when surprised, frightened, or excited. This tendency led to innovative hunting tactics, some of which are represented in ancient images. Throughout the Sahara, rock art includes ostriches. Some hunters trapped their prey in nets, as seen in this Libyan pictograph dating back 10,000 or so years. Others, including the Struthophangi hunters in Egypt and the Kalahari Bushmen in southern Africa, disguised themselves in ostrich skins and stalked them.[20] Ancient Egyptians, in the days of the chariot, apparently braced themselves against the sides of the vehicle to steady their aim as they fired arrows. The Neolithic etching from Libya shown here is modern in its artistic concept. The multiple heads are the etcher's shorthand for "flock."

Science

Ostriches are the largest extant land bird, although the extinct Great Elephant Bird (*Aepyornis maximus*) of Madagascar was larger still (one egg could accommodate the contents of seven ostrich eggs).[21] Male ostriches reach heights of nearly 8 1/2 feet (2.7 meters); females are somewhat shorter. Predation pressures and large size probably played a role in influencing the evolution of ostriches' cooperative breeding system.[22] That system allows several females to lay their eggs in the same nest, where only males and dominant females are permitted to incubate them. Numerous broods are later merged in order to provide for their care. These traits, ironically, favored human hunting success—and that pressure, coupled with climate changes that led to aridification of the ostrich habitat, nearly drove the birds into extinction. Some people have argued that had it not been for the domestication of ostriches, they would likely have become extinct by now. Wild populations in Egypt, for example, were last seen in 1991, suggesting their widespread vulnerability. Meanwhile, perhaps fortunately, ostrich farms have been established in Kenya, Egypt, Australia, New Zealand, the United States, and Argentina.

Hunters have sought skins, eggs, eggshells, meat, and plumes. Even gizzard contents were prized once it was discovered that the birds, on occasion, ingested diamonds along with the small pebbles used as grinding aids during digestion.

PLATE 12. *Pelicans from a Wall Painting in the Tomb of Horemheb* (no. 78) [untitled wall painting] (detail), Thebes, Egypt, early 1400s BCE

PLATE 12 **Pelicans and Their Eggs in Ancient Egypt**

Art

By the time the Sphinx was buried up to its head in sand, Horemheb was a royal scribe and a military general. Though not of royal blood, he became the last king of Egypt's Eighteenth Dynasty. He was buried in Thebes (c. 1350 BCE) in tomb KV57. This scene was painted on one of its walls. Art historians have pointed out that the chief fowler (the man who appears to be whistling to the pelicans, who are standing beside baskets of their eggs) roughly mimics the figure of the squatting man in the hieroglyphic (just beyond his right hand).[23]

Science

The Nile lies on a major migratory flyway, and the ancient Egyptians are thought to have had a close relationship with birds—both wild and domesticated. They were aware of seasonal migrations. They could differentiate between resident (nonmigratory) and migratory species, and they recorded various aspects of natural history. The domestication of birds there is a very old practice, one that even predates the dynasties, and ancient images show the use of herons as decoys to attract wetland birds into open traps, and the force-feeding of cranes and waterfowl. But the decision to feature pelicans (*Pelecanus* sp.) and their eggs in a tomb is puzzling. The pelican is said to have represented protection against snakes and safe passage after death, but since the chief fowler accompanies the pelicans here, perhaps they represent a valued resource for Horemheb in the underworld. If so, the image also suggests that pelicans were domesticated—which would be surprising.

Ducks, geese, pigeons, quails, and occasional chickens—all of which breed rapidly—were the principal domesticated fowl in ancient Egypt. It seems odd that the Egyptians would also choose to domesticate pelicans, which ordinarily lay only two eggs in each breeding season. But here is a fowler, apparently a person with an official function, superintending a group of pelicans. Pelicans are excellent fish catchers; perhaps some ancient Egyptians domesticated them for the same purpose that Japanese fishermen traditionally used Japanese Cormorants (*Phalacrocorax capillatus*), fitted with a restrictive collar, to catch fish. In fact, a pelican, a fish-clutching cormorant, and a group of fishermen hauling laden nets are included in an ancient Egyptian bas-relief (the *mastaba* of Mereruka), which was produced during the Sixth Dynasty (2345–2181 BCE), 1,000 years before Horemheb. In the bas-relief, neither bird has a collar, so perhaps fishermen merely watched them during foraging dives to see where to cast their nets. On the other hand, the baskets in this painting contain a lot of eggs, suggesting that they were a delicacy suitable for a pharaoh. Few pelicans are represented in Egyptian art, perhaps because they were more often admired and raised by commoners, thus keeping them off the walls of nobles' tombs. Why, then, are they here? Horemheb—unlike most Egyptian kings—was not of royal blood. Perhaps the pelicans are present as a special recognition of his history.

Ancient Science Art provides abundant room for speculation, but in the end it often leaves a mystery—in this case, one that perhaps future Egyptologists will solve. Some might see speculation as an inherent weakness of Science Art. We see it as an invitation to delve deeper into both the image and the science, to find out why the artist appears to have portrayed the shaggy-plumed Dalmatian Pelican (*Pelecanus crispus*), now a rare winter visitor to Egypt, rather than a White Pelican (*P. onocrotalus*) or a Pink-backed (*P. refescens*), which are also winter visitors; why grass appears below and above the eggs; whether two of the pelicans are preening or performing a courtship ritual; why priests were apparently forbidden to eat pelican meat; and, for that matter, whether the ancient Egyptians enjoyed eating the fishy-tasting meat and comparably fishy-tasting eggs.[24]

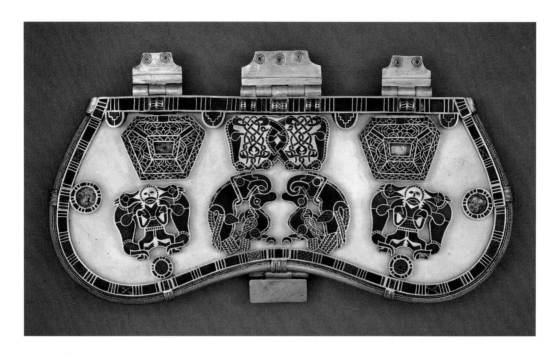

PLATES 13 and 13a. *Purse Lid from the Sutton Hoo Ship Burial* and detail, c. 600–650

PLATE 14. *Modified Detail of Purse Lid from the Sutton Hoo Ship Burial*, 1990/2007, by Darryl Wheye

PLATES 13, 13A, AND 14 **The Falcon as Hunting Assistant and Icon of Power**

Art

In 1939, just before World War II, an apparent gravesite was discovered amid fifteen mounds at Sutton Hoo on Britain's east coast. Excavations unearthed a 90-foot (27-meter) ship that had been dragged 600 yards (550 meters) from the water and buried beneath the sandy heathland. The ship is thought to be a tomb, with a burial chamber that included weapons, armor, coins, gold and garnet fittings, silver vessels, silver-mounted drinking horns and cups, and clothes. It also contained a leather purse with the lid in Plate 13—the richest jeweled lid of its kind yet found. The ship did not, however, hold human remains. Had there once been any? Probably. Analysis of a sample taken from the chamber indicated the presence of phosphate, suggesting that a body had been there but had completely decayed in the highly acidic conditions at the bottom of the ship. A few authorities question the analyses and think that the site was a memorial and that the ship symbolized the journey into the next life. Others, however, think that the site held one of the following kings: Eorthwald, who died in the late 620s; Raedwald, a powerful Christian king who lapsed into paganism and died in 624 or 625; or Sigebert or his brother Ecgric, co-regents, who were killed in 635 or 636. Of these, Raedwald or Sigebert are thought the more probable. The "who" is guesswork, but not the "when": coins found at the site date from 575 to 620.

Assuming that relics were placed in the pagan burial site to provision the king's afterlife, we can guess that they reflect his rank. Certainly they suit a life that was both lavish and barbaric. The relics include a wealth of gold and silver; surfaces are decorated with more than 4,000 hand-cut garnets and millefiori. The king's cloisonné-embellished purse, whose lid is seen here, was fastened with a gold buckle, and the base of its lid, now dissolved, was possibly whalebone ivory. But these relics also include a fierce-looking helmet covered with battle scenes and an apparatus described as a "scalp rack."[25]

Science

The raptor and duck predator-prey pairs decorating the lid were probably thought to confer strength and power, to memorialize the king's dedication to falconry, or both. As a symbol of political power, the predation narrative is reminiscent of that seen in a Greek coin produced around 412–411 BCE (see Plate 25). As a symbol of falconry, the pairs raise a question of timing. Falconry was indeed the sport of kings, and the first records confirming its practice appear around 2,000 BCE in central Asia or possibly western China. Within 500 years it was known in Asia Minor, and soon thereafter, in Greece. It was introduced to Japan from Korea by 244, and by the reign of King Ethelbert II of Kent (748–762) it was practiced in England, although, as this Saxon purse lid suggests, it might have been introduced considerably earlier.[26]

PLATE 15. *The Discreet Messenger*, undated, by François Boucher

PLATE 15 **Airmail in the 1700s**

Art

Is it odd that François Boucher, a French rococo artist, featured a message-bearing pigeon in a painting? Probably not. Madame de Pompadour, the mistress of Louis XV, was a major patron of Boucher's. When Boucher became Louis XV's chief painter, he worked at the royal residences at Versailles and Fontainebleau and documented court life. A pigeon—the cell phone of that day—could easily have been a feature there. Boucher's presence in this book, however, might have struck him as ironic, or it might perhaps have reassured him. He was known for using pastoral scenes to convey innocence and its pending loss. Bare feet were not unusual, but the decadence often associated with them is represented here in the stone carving; the woman conveys instead her uncommon literacy. Similarly, the cabbage, considered an antidote for drunkenness and often featured by Boucher, has no place here, and the typical caged bird, which stands for enforced propriety, is here liberated. Boucher had a sharp critic in the encyclopedist Denis Diderot, who said that Boucher's stereotypical coloring and artificiality arose from his failure to paint from nature. Boucher dismissed the criticism. The problem with nature, he said, was that it was "too green and badly lit."[27]

Science

Pigeon messenger service predates this painting by millennia—and continues to this day—but how much information a bird can carry has radically changed, for the leg ring seen here, which formerly held a handwritten message, can now hold microimages. Why use pigeons? Homing Pigeons, a breed of Rock Dove (*Columba livia*), will work in imperfect weather. They are fast fliers and are difficult to intercept, even for the occasional hungry Peregrine Falcon (*Falco peregrinus*).

In ancient Egypt, pigeon posts along the Nile apparently helped alert those downstream to the progress of the annual flood. The Assyrian military sent message-toting pigeons to relay intelligence information. Julius Caesar purportedly sent them home from Gaul, and Nero sent them, bearing the outcome of sporting events, to friends beyond Rome. The sultan of Baghdad established pigeon mail in 1150, and Genghis Khan set up an entire pigeon-mail network. Dutch sailors introduced pigeons to England, where stockbrokers and financiers apparently used them to get a jump on market news. Pigeons carried the announcement of Napoléon's defeat and, in 1879, news of the Siege of Paris. They carried messages in World War I, World War II, and the Korean War. Today, 1,000-mile (1,600-kilometer) flights for birds in the U.S. Signal Corps are routine; and a flight of 2,400 miles (3,860 kilometers) has been recorded.

Pigeon homing is also a seriously contested sport, and birds bred for their racing performance in these competitions, which can span hundreds of miles from release point to home target, can bring a high price from aficionados. The performance itself has long fascinated students of animal orientation and navigation. The complex mixture of sensory abilities that the birds apparently utilize includes sensitivity to discontinuities in the earth's magnetic field, orientation to the direction of the sun's movement, and memory of visual landmarks or knowledge of patterns of wind currents.[28]

PLATE 16. *Happy Thanksgiving*, 2000, by Sally M. Berner

PLATE 16 **Thanksgiving Turkeys: The Dark Side**

Art

Sally Berner's well-received painting of turkeys says a lot about our dependence on avian resources, our responsibility to see that they are properly managed, and our tendency to overlook what makes us uncomfortable. She herself notes a schoolgirl's remark: "Oh, that's good, they got the license number of the truck." The painting also says a lot about the importance of finding as many venues as possible for Science Art. This painting was juried into the prestigious *Birds in Art* exhibition at the Leigh Yawkey Woodson Art Museum, published in its annual catalogue, and included in the exhibition's national tour. It also won first prize (out of 12,000 entries) in the "Animals" category of *The Artist's Magazine's* annual art competition and was featured on its Web site.[29]

The artist's caption for the *Birds in Art* catalogue explains the genesis of the painting: "Last September, while driving through West Virginia, I had my first encounter with a truckload of turkeys on their way to market. The dismal weather seemed to reflect the plight of the turkeys, and I found the scene particularly disturbing. I took reference photographs so I could do something to bring attention to the inhumane treatment of these birds headed for our Thanksgiving tables. I usually paint photo-realistic animal portraits, but here I used an impressionistic style to elicit a more emotional response."[30]

Science

A Science Art caption for this painting might also provide information on turkey production. For example, U.S. turkey consumption climbed from 8.1 pounds (3.7 kilograms) per person in 1970 to 17.5 pounds (7.9 kilograms) per person in 2001. This translates into 5.5 billion pounds (2.5 billion kilograms) of turkey for domestic use that year, the year after this painting was made. An additional 200,000 metric tons (nearly 220,500 short tons) of turkey products were exported to major foreign markets.[31] To assess the full impact of the industry, however, information about resource use in rearing, processing, and transporting the birds is needed as well.

As concern about the spread of bird flu continues to generate information about the poultry industry, consumers will be in a better position to understand the role of this major food item in the American diet. Had Benjamin Franklin succeeded in having the Wild Turkey (*Meleagris gallopavo*) designated as the national bird, the industry arising around this representative of the poultry family might have been quite different.

ROOM 3

Birds as Teaching Tools

The pictorial record is an archive of evidence. It shows us, for example, how changes in a species' range, as recorded in historical images, can indicate changes in climate, or the inroads of an invading species, or the advance of habitat degradation caused by the expansion of human activities. It shows us how artists have helped us understand the intricacies of bird anatomy and the extent of bird diversity and how they remind us specifically of the role that birds have played in helping us build up a picture of evolution. The often-illustrated Galápagos finches, for example, became an icon of the evolutionary process. (By evolving bills that differed in size and shape, Galápagos finches were able to specialize in different foods and occupy different ecological niches. Some forms became stout-billed and could crack large, hard seeds; other forms became scythe-billed and could sip nectar from tubular flowers. A handful of other forms came to specialize in foods such as bark-dwelling insects or mammalian blood.)

Artists also remind us of the role that birds have played as ancient models for replicating flight, inspiring both those hoping to use feathered wings to fly, like Daedalus (see Plate 21), and those hoping to devise mechanical ones, like Leonardo da Vinci (see Plate 43). Aerial images remind us that birds have also served as models for actual flight. For example, after biologists successfully raised a generation of cranes in a portion of their range where the species had been wiped out, they had to teach them how to reach their wintering grounds. In a man-bites-dog reversal, the artist-turned-biologist William Lishman flew his ultralight aircraft trailed by a flock of reintroduced birds from their breeding grounds to their wintering grounds. He began training with a flock of geese, moved on to Sandhill Cranes (*Grus canadensis*), and finally led a flock of highly endangered Whooping Cranes (*Grus americana*) from their breeding grounds in Wisconsin to their wintering habitat in Florida.[32]

And artists remind us of the role that birds have played in teaching us to recognize changes in environmental quality. In the 1960s they portrayed species like Brown Pelicans (*Pelecanus occidentalis*) that were being exposed to enough DDT in their diet to interfere with calcium deposition, a key component of eggshells. These gangly birds broke their eggs just by sitting on them and became indicators—in fact, icons—of the dangers associated with the indiscriminant use of pesticides. Now, decades after DDT-use was banned in North America, artists are chronicling the way the birds have rebounded. Artists have also recorded the role that birds played in teaching our ancestors to notice changes in weather conditions, perhaps allowing them to forecast certain kinds of weather in advance of modern devices. Taken together, birds and artists have informed and inspired generations of researchers and other viewers of their work.

PLATE 17. *Auk in Cosquer Cave,* Alpes Maritimes, France, c. 17,000–16,000 BCE

PLATE 18. *Modified Detail of Auk in Cosquer Cave,* Alpes Maritimes, France, 2005/2007, by Darryl Wheye

PLATES 17 AND 18 **Mediterranean Auks in an Underwater Cave**

Art

In 1985, while deep-sea diving 110 feet (35 meters) below the surface at Cape Morgiou, near Marseilles, France, Henri Cosquer discovered the small entrance to a cave. Venturing 450 feet (135 meters) inside, he discovered an expansive, air-filled chamber. When he returned six years later, he discovered the chamber's gallery of Paleolithic cave art, including this outline of what appears to be a Great Auk (*Pinguinus impennis*), painted with charcoal on a vaulted ceiling not far from two smaller images that also appear to be Great Auks.

Science

The underwater entrance presented two obvious questions. What was this decorated cave doing in deep water? And, could the pictures in its gallery possibly be authentic? The story that emerges from subsequent research is as remarkable as the art. Assessments of the tiny, slow-growing calcite crystals coating the images established that the images had been painted at a much earlier time. When the Paleolithic artists painted in the cave, it appears that they were working on a cliff some 250 feet (75 meters) *above* the Mediterranean shoreline and several miles inland. The dramatic shift in the cave's elevation resulted from climate changes affecting sea level; changes over the past 30,000 years have shifted the location of the cave entrance from above to below sea level. Researchers have taken samples from charcoal drawings in the cave for radiocarbon dating. These tests apparently confirmed two major production phases: one about 27,000 years ago, when finger tracings and stenciled hands were created; and a second phase between 18,000 and 19,000 years ago, when most of the animals, including this one, were completed.[33] The second production phase came shortly after the last glacial maximum, when much of the Earth's water was tied up in great ice sheets, and sea levels were lower than today's by 300 feet (90 meters) or more. The cave's entrance would have been high and dry at that time, but progressive warming over the past 10,000 years—during the Holocene period—eventually immersed the entrance.

The depictions of marine life in Cosquer Cave suggest the possible value of particular organisms to our forebears. For example, numerous seals are shown, often speared. Evidence of hunting is not limited to seals—28 percent of the animal images include arrows or spears. The Great Auk was surely known to the Paleolithic residents of the region. The 29-inch (74-centimeter) flightless bird, also known as the Northern Penguin, became extinct in its modern northern range in 1844 following several generations of ruinous overharvesting. It had been an extraordinarily effective diver, like some of its smaller relatives, like Razorbills (*Alca torda*) and Puffins (*Fratercula arctica*), that are still found in the region today. But the Great Auk could well have been called the Great Awkward for its clumsy gait on land, which made it easy prey for hunters. One of the clues identifying this bird is its stunted wings, which, though usually held clamped tightly to the body, were apparently held "a little out (so that light shows under it) when it began to run" (Plate 18).[34] In those colder times the auk occurred in France and the Mediterranean and probably dove within sight of the historic cave found by Cosquer.[35]

This painting not only indicates the species' former distribution in the region but also suggests that it had special value to the artist and to the local human community. Perhaps it was also valued as food, but we know that early hunters in historic times took advantage of the extremely dense oil and fat layers that protect the Great Auk against icy water, lipids that made the Great Auk inflammable. There is clear evidence that nineteenth-century hunters burned piles of their carcasses as fuel.[36] Also, more than 130 natural and modified limestone lamps have been found at Lascaux, along with torches, spare wicks, and flints, suggesting the possibility that our Cosquer Cave artists may have used these flat, fat-burning lamps, too, and fueled them with auk oil.[37]

PLATE 19 *(above). Animals Sketched from Nature (Xiesheng Zhenqin Tu)*, undated, by Huang Quan

PLATE 20 *(right). Page from a Monk's Drawing Book* (pl. 11b), England, 1300s

PLATES 19 AND 20 **Teaching Models in China and Europe**

Art

In China, flower-and-bird painting developed into an independent category of art during the Tang Dynasty (618–907) and reached a pinnacle during the Five Dynasties (907–960), when Huang Quan was active. Huang often sketched rare birds directly from specimens, and the standards he set lasted a century.

Huang is sometimes credited with originating a method of painting that uses colors in graded washes to build up forms instead of relying on line to delineate forms. This image (Plate 19) is the only one of his surviving paintings whose authenticity is uncontested. Two well-known stories convey Huang's talent. In one, he painted a set of six cranes that were so realistic that a real crane stood beside them, leading the emperor to rename the hall housing the paintings the Hall of the Six Cranes. In another story, a live eagle spotted a pheasant he had painted and tried to seize it. The painting seen here includes a cochoa (an Old World flycatcher), a tit, a Tree Sparrow (*Passer montanus*), a wagtail, and a Northern Wheatear (*Oenanthe oenanthe*). Huang may have selected the wheatear because it was common and was seen without difficulty. (Today the species is a model for migration studies. Its breeding range covers almost half the world, and this easily viewed open-country bird is well suited for studies of the comparative roles of instinct and adaptation in long-distance migration that are key for understanding migrants at risk from climate change.)[38]

In contrast, it is doubtful that the birds in the monk's drawing book were drawn from life. Nor, it is hoped, was the rather devilish-looking bat-like specimen that appears at the top of the page shown here (Plate 20). Some birds appearing on the page can be identified more convincingly than others. Beginning at the upper right corner, these include a Green Woodpecker (*Picus viridis*), a Crane *(Grus grus)*, a gull, a Spoonbill (*Platalea leucorodia*), a raptor, and a Skylark (*Alauda arvensis*). Among the mysterious rest could be a cuckoo, a nightingale, a nightjar, and a finch. This page in the notebook is one of eight that feature birds, and the notebook is the only medieval English model book to survive.[39]

Science

The study of ten bird species by Huang Quan served as a model; that is, the artist is said to have produced it to teach his son to draw. The birds portrayed are more accurately represented than those in the monk's drawing book, produced 400 years later, which was also used as a model for copyists. The disparity between these models probably has less to do with the talent of each artist than with differences in their understanding of underlying behavior, ecology, and comparative anatomy. Repeated inaccuracies appear in medieval images. The tradition of using model books introduced or perpetuated errors as artists copied inexact models, or copied copies of those models, rather than basing their paintings on live birds or specimens. Like a genetic line passing along a dysfunctional trait, the monk's drawing book contains flaws that appear in surviving medieval artwork. The gull seen here, for example, can also be seen in the thirteenth-century *Tenison Psalter;* and the hawk and duck can be seen in *Materia Mecia,* a north Italian manuscript from the same period.

The Chinese and European plates highlight some of the benefits and limitations of extrapolating science from art. Like a stratum of rock containing fossils, the painting by Huang provides a datable record of particular species seen in a particular region at a particular time. In contrast, in the page from the monk's drawing book, where some species can be identified but others cannot, there is less science about the birds to extrapolate but possibly more information about the opportunities available to medieval monks to view birds in the course of their daily work.

PLATE 21. *The Art of Flight, Daedalus (Il Volo di Dedalo)*, from the Bell Tower (Campanile), Florence, Italy, c. 1334–1348, by Andrea Pisano

PLATE 21 **Daedalus and Four-Winged Flight**

Art

This carved figure is from a church in Florence, Italy, but it is not an angel with abnormally large wings. It depicts Daedalus—Greek for "the skilled one"—who made wings of feathers and wax so that he and his son, Icarus, could escape their confinement on Crete. In the story, when Icarus flew too near the sun, the wax melted, and he fell into the sea and drowned. The marble carving is part of a series that wraps around the campanile. The set of carvings illustrates human quests, an atypically secular subject for a church, but one that reflects the rising authority of universities. More particularly, the series portrays the variety of moral and intellectual influences present in mid-fourteenth-century Europe. The section that contains Daedalus illustrates the state of affairs following the expulsion from Eden and features those who worked to improve life outside the garden.[40]

How, we might ask, could a human being's attempt to fly be viewed as a realistic effort to improve the human condition in the 1300s? To some, replicating flight was doubtless a dramatic demonstration of human arrogance. For others, it apparently testified to human potential. Over the centuries, people in the latter group found progress toward flight painfully slow. Time and again, efforts to design and construct airworthy apparatus—like those made by the medieval engineers Roger Bacon and Leonardo da Vinci—failed.[41] A small tablet on a hill above Fiesole, overlooking Florence, marks the edge of the cliff from which Leonardo's flying machine took off. It honors the designer but pays scant attention to the poor apprentice (today one might read "graduate student") who crashed and suffered serious, possibly mortal injury in the attempt to achieve flight.

Science

The physiological requirements for flight include a strong, light framework that houses a respiratory system more efficient than that of mammals, a circulatory system able to sustain muscles taxed by long, strenuous exertion or rapid bursts of speed, and a highly tuned nervous system responsive to instantaneous changes in conditions. Donning a pair of wings and flapping them quickly will not make mammals airworthy. But after centuries of perseverance, the development of ultralight materials, and some complex mathematical calculations, the myth of Daedalus's flight became reality (see Plate 43).[42]

Archaeopteryx, a 150-million-year-old transitional form linking reptiles and birds was probably unknown to Andrea Pisano, who sculpted the Daedalus relief. This precursor to modern birds had specialized hind-limb plumage resembling flight feathers that apparently allowed these limbs to serve as airfoils. These "rear wings" would have made *Archaeopteryx* capable of slowing down and making sharp turns, which should have reduced collisions and some of the risk during landings. The evolutionary switch from four wings to two might have improved running, swimming, or grasping prey.

Examining the fossil evidence for four-winged flight has had a history. In 1915, William Beebe considered a four-winged stage in avian evolution after looking at the Berlin specimen of *Archaeopteryx*. Recently, discoveries of fossils having well-defined hind-limb feathers have led researchers to revisit *Archaeopteryx*. In 2004, a year after X. Xu and colleagues suggested the possibility of a four-winged "tetraplex" with respect to *Microraptor gui*, Per Christiansen and Neils Bonde took a fresh look at the Berlin specimen, finding that the feathers on the back and legs were, in fact, vaned but were smaller than flight feathers, and those on the neck resembled hairlike protofeathers. And in 2005, Nick Longrich again called attention to the hind-limb feathers in *Archaeopteryx*. Two others, F. Zhang and Z. Zhou, also concluded that leg feathers in newly discovered early Cretaceous specimens in China had aerodynamic qualities. These recent episodes have made it difficult to overlook the evidence of hind-limb plumage in *Archaeopteryx*. Viewers will note the "hind limb" plumage in Pisano's rendering of Daedalus made 700-plus years ago.[43]

PLATE 22. *Portrait of Balthazar Sage*, 1777, by Colson (Jean-François Gilles)

PLATE 22 **Animal Testing for Toxic Gases in the 1700s**

Art

Those who find Colson's portrait a less than great painting and his subject a less than great scientist may find the image memorable nonetheless. Its artist was familiar with engineering, architecture, and sculpture, wrote works on perspective, exhibited at various salons, and toward the end of his life was elected to both the Academy of Sciences and the Academy of Arts and Letters of Dijon, France. He was also a portraitist whose clients included musicians, comedians, members of the military, and, as seen here, the pharmacologist Balthazar Sage (1740–1824).

When Sage was twenty, he began giving public lectures at his family's pharmacy on both chemistry and assaying techniques. He also began building what would become one of the largest and most important mineralogical collections in Paris. Sage did not prove to be a particularly competent chemist, but his mineral collection positioned him to establish the Paris School of Mines, to which he ceded his collection for an annual income of 5,000 livres. His contacts with the French court led to his arrest during the French Revolution, and he was released from prison only after paying a hefty fine. When he died, blind and impoverished, he was still a staunch royalist, but he had been a scientist in only a vague sense of the word.[44]

Science

Colson painted the portrait when Sage was teaching at the new School of Pharmacy. He shows Sage seated beside an oven used during demonstrations. Two of three songbirds appear to be dead, apparently after exposure to a toxic agent. Unlike the expensive cockatoo featured in Joseph Wright's *An Experiment on a Bird in the Air Pump* (see Plate 44), these three birds were probably local wild birds and considered expendable.[45]

This painting allows us to connect the age-old use of the canary in the coal mine with the use of birds to find warning signs of environmental vulnerability. Who better to bring into the laboratory a bird to indicate the buildup of toxic gases than Sage, the founder of the Paris School of Mines?

PLATE 23. *Ahead of the Storm*, 1989, by Lee Stroncek

PLATE 23 **Forecasting Tomorrow's Weather**

Art
These Snow Geese (*Chen caerulescens*) are flying right at the storm front. The Montana artist Lee Stroncek wrote of his painting, "I have often heard and sometimes seen geese in the late fall, often at night, running ahead of a major snowstorm." Does this mean that nocturnal migration might forecast changes in the weather? The short answer is maybe. Migrating birds indicate both a need to travel and conditions that allow it, but travel speed, altitude, and storm size factor in. Geese fly at altitudes of about 1,000 feet (c. 300 meters) to about 3,300 feet (c. 1,000 meters), with a few flocks reaching 4,300 feet (c. 1,300 meters), depending on visibility and wind, and the birds often sustain speeds of more than 30 miles per hour (c. 50 kilometers per hour) for extended periods.

Using geese to predict weather has a long history. Ancient Romans, for example, apparently measured the breastbones of geese that had hatched the previous spring to predict the severity of the upcoming winter: a thin bone indicated mild weather; a thick one, "heavy" weather; and a translucent one with light spots, wet weather.[46]

Science
Geese use the sustained winds of weather systems during migration. In North America, where weather typically moves eastward, they use trailing high-pressure systems during fall migration and the west side of high-pressure systems, or occasionally the east side of low-pressure systems, in the spring. How important is flying in V-formation to outpacing a storm? Anyone who attends swim meets or watches bicycle races knows that participants regularly save energy by taking advantage of the upwash field generated by adjacent competitors or by "drafting" on the rider ahead. Formation-flying birds take advantage of the lift provided by air movements that result from the wing-beat cycle of the bird ahead, which may allow the entire flock to extend its flying range by saving energy. Investigations of lift at NASA's Dryden Flight Research Center using two F/A-18 aircraft, for example, show a reduction in drag that suggests the possibility for significant fuel savings. The boost may prove increasingly significant, since global climate change is modifying the timing of migration, as it has geographic range and all levels of community ecology. One study of the Canadian Arctic found that Snow Geese and Canada Geese (*Branta canadensis*) populations advanced breeding by a full month between 1951 and 1986. Analysis of climate change over the past three decades by C. D. Thomas and associates showed changes in species distributions and abundance that led them to project extinction for 15 percent to 37 percent of all species over the next half-century, depending on which climate models prove most predictive.[47]

ROOM 4

Birds as a Means of Understanding Biology

The images in this room present aspects of bird biology that artists have recorded, although this is such an extraordinarily rich and productive area of study that it is impossible to derive a sample that represents its full scope. Our choices were not dictated by taxonomy, or evolution, or particular bird behaviors. Rather, we made selections based on three general themes—the history of art, human interest in birds, and birds and nature—and conforming to the general dates of production used in the other four rooms of the Lower Gallery.

Birds have played an essential role in helping humans to decode nature. Characteristics that are particular to birds, especially feathers, flight, and behaviors associated with singing, have been recorded ever since a cave wall was seen as a suitable canvas, if not earlier. Artists have provided a visual primer on feathers. They have shown birds close-up, so we can see individual overlapping feathers still faintly reminiscent of the reptilian scales from which they evolved. Artists have shown how bright coloration attracts mates and predators, how cryptic coloration can fool the latter, and how habitat and lighting sometimes allow even bright colors, like those of yellow birds perched in dappled light, to become cryptic. They have shown thermoregulating birds as they stand on one leg with the other tucked into insulating feathers, and preening birds as they oil and zip ruffled feathers back into place.[48]

Artists have also provided a visual primer on flight. They have shown how wings split the air and how briskly birds flap them for takeoff and how deeply birds arch their wings for landing. They have shown how wing shape and wing slant relative to airflow allow air to move more rapidly over the upper surface than beneath the lower one, a difference that reduces the pressure on the upper surface, sucking the wing upward and creating lift. They have shown how differences in air temperature and current affect flight, how columns of warm thermals support broad-winged raptors and vultures as they soar high over land, and how the layer of relatively calm air hanging just above the ground or water attracts birds, allowing them to stay aloft using less energy than if they were in more turbulent air. Artists have called attention to the serrated, sound-dampening primaries of hunting owls and the dive-enhancing torpedo-shape of flightless penguins. They have made it easier for viewers to assess whether geese flying in V-formation might be saving energy and to see some of the cues that birds use to orient and navigate on short flights and on long annual migrations.

And artists have provided a visual primer on displays and moments of song production. Examples include allegorical narratives featuring iconic birds, like the Nightingale (*Luscinia magarhynochos*), but artists have also portrayed songless birds, like the Common Raven (*Corvus corax*), that manage instead with the aid of a vast repertoire of calls (see Plate 51). Three hundred years ago, people used musical notation to reconstruct bird song visually. Today researchers use sound spectrographs and sonograms.

Artists have also portrayed elements of community ecology that show avian relationships with local plants and animals, or show habitat partitioning that, for example, places longer-legged aquatic birds in deeper pools, or demonstrate the effects of invasive species. They have portrayed dominance displays that reveal pecking order and territorial boundaries, and assaults by small birds on large avian predators who enter into their breeding areas (mobbing; see Plate 53). They have portrayed mating strategies and courtship; nesting and parental care; methods of foraging and hunting—from strenuous aerial pursuits, to cruising for carrion, to kleptoparasitism (as when a Bald Eagle [*Haliaeetus leucocephalus*] harasses an Osprey [*Pandion haliaetus*] into surrendering its fish).

Artists are increasingly assisted by new technologies—in particular, high-speed photographic imaging using strobe-light illumination, which captures quick movements and permits new levels of behavioral analysis. And artists are increasingly assisted by new research on birds that is widely available online and allows them to portray new findings in, for example, the dual processes of speciation (the splitting of species into new forms) and hybridization.

PLATE 24. *Owls in Les Troix Frères Cave*, Ariège, France, c. 30,000–17,000 BCE

PLATE 24 **Nesting Owls and the Classification of Paleolithic Art**

Art

Two birds and their chick in this etched outline, which has been in the Les Trois Frères (Three Brothers) Cave in the French Pyrenees for at least 17,000 years, are traditionally identified as a family of Snowy Owls (*Nyctea scandiaca*). Once again, questions arise when viewing Paleolithic owl images: "Why owls?" and "Why here?" The bones of Snowy Owls have been found in various caves, and more than 1,100 bones from more than eighty large owls have been found in association with Paleolithic tools. The Les Trois Frères Cave also houses the image of an apparent owl-headed human figure with antlers, paws, and a tail. Such chimeras are sometimes interpreted as shamans in a trance and sometimes as hunters disguised to approach their prey, but no one knows for certain what they represent (see also Plate 31).[49]

Although Paleolithic chimeras remain puzzling, art scholars have reached agreement about some aspects of cave art, finding similarities among images that convey motion and standardizations in size, color, height above the cave floor, and position. There is less agreement, however, about how to classify the assembly of art in any given cave. The Paleolithic art scholar L'Abbé Henri Breuil (1877–1961) chose not to classify individual works as part of a coordinated series, but André Leroi-Gourhan (1911–1986) believed that the artistic content of each cave could be organized into an integrated composite. He also devised a classification of Paleolithic images into four stages of complexity and categorized elements by symbolic meaning (for example, branching signs as male; wounds as female). But the discovery of Chauvet Cave made the Leroi-Gourham system obsolete. Radiocarbon dating and accelerator mass spectrometry (AMS) give a relatively secure date of 30,000 BCE for the Chauvet Cave art. In addition to housing some of the oldest Paleolithic paintings on record, Chauvet Cave houses some of the most sophisticated.[50]

Science

The biologist Stephen J. Gould (1941–2002) made some helpful comparisons between biological evolution and the cultural evolution of style in art. He wrote that from the perspective of human evolution, the standard view of art as a system that is refined by time and progresses from crude or simple to elegant or complex is anti-Darwinian. In a more Darwinian view, changes in art would be seen as adaptations to changes in local conditions, as accommodations to the adoption and transmission of new stylistic elements, not as developments in a march of progress. This evolutionary view of art may become clearer if we stand back to view and assess the entire human timeline. The earliest inhabitants of the regions of Europe where cave paintings have been preserved might have left Africa 45,000–60,000 years ago. The earliest cave artists began leaving records around 30,000 years ago. That makes the artists much closer to us in time than to the first modern humans, who arose in Africa some 150,000 years ago. It also makes Paleolithic art comparatively modern, and the "art progresses" view somewhat shortsighted. After all, some art produced today can be mistaken for images produced 30,000 years ago—and vice versa.

Taking a somewhat different approach, R. Dale Guthrie argues in a recent work that much Paleolithic art was probably created by women and children or inexperienced artists, but the most realistic animal images would have been produced by hunters (all men, as far as we know), who were dependent on their knowledge of the animals to find and kill them. Were the owls in Les Trois Frères Cave particularly valued by hunters? Guthrie notes that Eskimos in Alaska traditionally hunted owls as a delicacy. In addition, fletched feathers used to stabilize spears (the same way they are used to keep arrows flying straight as they lose speed) are recorded in the cave.[51]

Alternatively, Snowy Owl families could represent devoted parental care. Females brood their young full-time, males provide all the food, and families stay together at least until the autumn. Egg laying starts again in the spring. Was this strategy a model for our ancestors?

PLATE 25. *Ten Drachma Silver Coin*, Arkagas, Sicily, c. 412–411 BCE

PLATE 26. *Modified Detail of Ten Drachma Silver Coin*, Arkagas, Sicily, 1990/2007, by Darryl Wheye

PLATES 25 AND 26 **Predators and Prey on Ancient Greek Money**

Art

Greek coins first appeared about 600 BCE and within a century had spread well beyond the mainland. Although the Greeks had no banks, from an economic standpoint the effort to standardize payments was revolutionary. From an artistic one, minting coins with images like the eagles seen here established a standard that endures to this day. The eagles conveyed a message to the citizenry, just as those on modern coins do now. Then, however, the birds were considered to be emissaries of Zeus, serving as omens and handing out punishment. Zeus had one eagle do to Prometheus more or less what an eagle appears to be doing to the hare in the coin. The circadian punishment—the eagle extracted Prometheus's liver each day (the liver repaired itself each night)—was imposed for two acts of defiance. In the first he shorted Zeus a measure of sacrificial meat by providing a choice between two offerings: one contained ox entrails wrapped in fat, and the other contained choice bits of ox placed in its stomach. Zeus made a bad choice and in his rage took fire away from humanity. In the second act of defiance Prometheus returned fire to humanity, and in a second display of rage Zeus had Prometheus arrested and chained to a rock for the daily liver extractions.

Eagles often appear on coins from Akragas (present-day Agrigento), so perhaps the image was a coat of arms. Still extant in Agrigento are the remains of a large temple dedicated to Zeus, built during the reign of Theron (488–472 BCE), more than half a century before this coin was minted.[52]

Science

In ancient Greece the eagle was a symbol of victory, freedom and Zeus. Two eagles feeding on a pregnant hare constituted a powerful metaphor about overtaking a vulnerable but rapidly growing community; the biological message is clear even though it reflects a sophisticated ecological principle. The hare is an animal associated in contemporary as well as ancient times with "irruptive" population cycles: boom years with high birthrates, are followed by low-density "bust" years as resources become sparse from overconsumption. A significant increase in the population of hares (the prey) will lead to a significant increase in the population of eagles (the predators). But the balance between predator and prey is self-regulating. If the prey population overwhelms its resources or invites disease, it will crash, to be followed by a sharp reduction in an increasingly hungry predator population.[53]

When the Greek dramatist Aeschylus used the metaphor in his play *Agamemnon* (458 BCE), it served as a dramatic reminder that military success always bears the hidden cost of future retribution. The play recounted the triumph of the Greek warrior princes Agamemnon and Menelaus over the Trojans. Priam, the king of Troy, reputedly sired fifty sons and fifty daughters—a crowded lineage and a threatening rate of growth.[54] The image on the coin, then, warned of the dual threats of overpopulation and reprisal. It is a curious coincidence that a grasshopper, another species whose populations periodically overwhelm its food supply, is also depicted here.

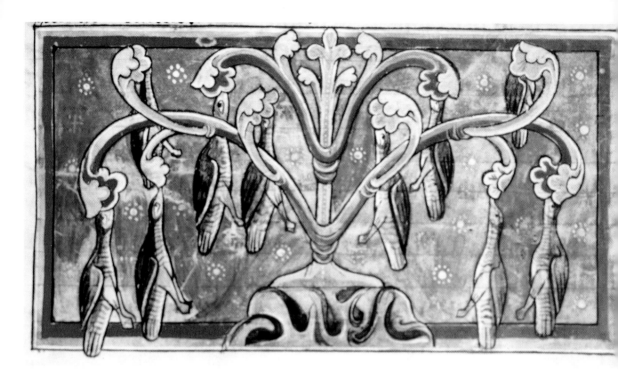

PLATE 27. *Barnacle Geese,* in the *Harley Manuscript,* England, c. 1230–1240

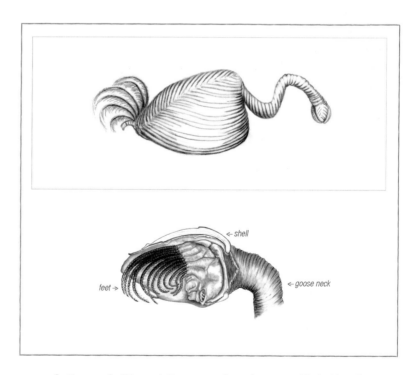

PLATE 28. *Goose-necked Barnacle* [*Lepas* sp.], after a drawing published by Ulisse Aldrovandi, c. 1639, and *Goose-necked Barnacle Cutaway,* after a drawing by Lois Rein, c. 1986, both by Darryl Wheye, 1996/2007

PLATES 27 AND 28 **The Goose-necked Barnacle and the Barnacle Goose**

Art

In Plate 27, Barnacle Geese (*Branta leucopsis*) are shown hanging from a tree, like some kind of unexpected fruit. Was this picture in the 775-year-old *Harley Manuscript* a metaphor? Was it a Gothic joke? Apparently it was neither. The breeding grounds of these geese lie far to the north, and Europeans had yet to find them, so myths arose to explain why the birds always materialized fully formed each fall. Some inspired minds concluded that they developed on trees overhanging bodies of water and, upon reaching maturity, fell like ripe fruit safely onto the surface and paddled away. Others concluded that the birds hatched from stalked barnacles adhering to ships, driftwood, or floating tree trunks.

The reason behind the apparent foolishness of this image has much to do with religious custom. Since the birds could be thought of as hatching from trees or from barnacles, it was argued, their flesh was not meat. This suited the eleventh-century cardinal bishop of Ostia, Pietro Damiani, who proclaimed that the flesh of the Barnacle Goose was not meat and therefore was acceptable protein during Lent and other meat-fasting days. Alexander Neckam wrote in *De naturis rerum*—a book on natural history crafted with a moral filter—that since the birds were produced from water rather than from mating, people could eat them freely during Lent, as they do fish.[55] Neckam was abbot of the Augustinian monks at Cirencester and stepbrother of Richard the Lion-Hearted, so his words carried some weight. A century later the idea was still circulating in Ireland, and although the great compiler Albertus Magnus (c. 1193–1279) refuted it, images like this one were still being produced another 250 years after that. Even the anatomist Ulisse Aldrovandi (1522–1605) professor at the University of Bologna, published a picture that emphasized the visual similarities between the barnacle and the goose (Plate 28). Thus, the picture in the *Harley Manuscript,* like others that followed, kept the myth alive and apparently kept the geese on the table during meat-free holidays.

Science

Why the Europeans associate the goose and barnacle to the point of calling one the Barnacle Goose and the other the Goose-necked Barnacle (*Lepas* sp.) seems to be a case of unrelated organisms that share similar form. The Goose-necked Barnacle secretes adhesive that attaches its antennae-free, gooseneck-like head to wood or other substrates. Its six pairs of featherlike legs extend through the opening in its shell plates and propel food toward its mouth. The feathery appendages were mistaken for incipient plumage. Interestingly, although these barnacles appeared for centuries in illustrations of ponds and lakes, they live exclusively in marine habitats. Eventually the lore faded into obscurity, but the Barnacle Goose held onto its name.[56]

PLATE 29. *The Parliament of Birds*, one of a pair of undated paintings, by Carl Wilhelm de Hamilton

PLATE 29 Visual Primer on Avian Diversity

Art

In 1177, Farid ud-Din Attar wrote *Manteq at-Tair* (*The Conference of the Birds*), a poem constructed around a pilgrimage, which moves from the crowded, random everyday world toward a higher plane, allowing the author to comment on contemporary society. As the pilgrimage proceeds, much of the text involves the responses of a Hoopoe (*Upupa epops*), an Old World bird, to the questions and the protests of other birds. This is not to be confused with Chaucer's 700-line poem, *The Parliament of Birds,* that appeared two centuries later (c. 1380). Chaucer's poem is constructed around thirty-six species that meet each Saint Valentine's Day to select mates. Carl de Hamilton's painting, produced 300 years later, is based on Chaucer's poem.[57]

In the Chaucer poem a bookish narrator reads the description of a mystical experience entitled "Scipio's Dream" found at the end of Cicero's 2,000-year-old classic, *On the Republic* (51 BCE). When the narrator finishes reading, he acknowledges that he is unsatisfied and hungry for what he lacks. He falls asleep and dreams that Scipio's father, Africanus, pushes him into an idealized garden, where he encounters numerous characters, including Lady Nature, who is overseeing the annual parliament of birds. She gives each bird a mate but fosters harmony by also allowing choice. Eventually the cacophonous flock flies away, and Chaucer has his narrator awaken from his dream. In the end the narrator fails to take up the challenge to experience life firsthand, promising instead to read books that induce even better dreams. Now, six and a quarter centuries later, birders would be the first to reaffirm the benefits of remaining attuned to the world of birds, although avian mating systems have generally proved to be anything but the orderly and simple parliament that Chaucer describes. It might be tempting to argue that the lek mating system used by some members of the grouse family is an exception. As we will see, male grouse display at a communal arena that is used year after year, but the assembled birds fight to maintain position (Plates 68 and 69).[58]

Science

In his painting of the parliament of birds, de Hamilton made each species identifiable and increased the number from the thirty-six mentioned in the poem to more than sixty; they are listed in the note.[59] De Hamilton, who was born in Brussels, did not, however, include the total number of species known in his region in his day. Species lists made hundreds of years earlier included 114 European birds. (There were hundreds more than that.) Although those old lists, like the ones compiled by Albertus Magnus of Germany (c. 1193–1279) and Frederick II of Prussia (1194–1250), were ornithological milestones, only paintings like this one by de Hamilton, or expensive books with plates or specimen collections, even if not comprehensive, were visual primers of the existing diversity.

Bird studies changed forever when lists and images were finally combined into illustrated field guides. The field guides helped put to rest scores of misunderstandings perpetuated in lore and literature. For example, it could be claimed with justification that among members of the same species, bird vocalizations encourage orderly sociability by drawing mates to one another, establish and strengthen pair bonds, entice birds to feed, and threaten outsiders. But mixed-species aggregations are definitely not mate-selecting "parliaments"; where they exist, they allow for cooperative feeding and defense against predators, or reciprocal buffer against cold temperatures and other extreme conditions.

By 1380 the Medieval Warm Period (800 until 1300) had already passed, so February 14 was rather early in the British year for mate selection. We now have to consider the unhappy possibility that global climate change could eventually bring that date within an appropriate range.

PLATE 30. *The Enchanted Domain I (Le Domaine Enchanté I)*, 1953, by René Magritte

PLATE 30 Pigeon-Plant Coevolution: Flesh Is Part Grass

Art
Unlike traditional images of bird-related chimeras—the sphinx, the griffin, the winged steed, the dragon, and the angel, all of which usually signified speed or spirit—today's chimeras sometimes represent real flesh-and-blood constructions and other things as well. That is the case with René Magritte's seed-producing plant that is also seed-eating birds. Magritte, an academy-trained Belgian surrealist painter, often placed familiar objects in unfamiliar settings or combined objects in unsettling ways (for example, by inverting their relative sizes).[60] Birds are not uncommon in his works.

This painting, Magritte's largest work, is part of a mural that spans the walls of a casino. (A team of artists copied his original work, projected onto the walls with slides.) The mural, with its lesson of coevolution, illustrates the educational possibilities afforded by Science Art when placed in public spaces. The placement in a casino could be seen as apt. In various casino card games players "feed" their hands with cards from the "house," behaving somewhat like foraging birds.

Science
Why Magritte might have chosen pigeons in this image becomes a bit clearer if we consider the effect of the birds' foraging behavior on seed-producing plants. They serve as active selective agents as they forage, determining which seeds are eaten and removed from the population, along with their genes, and which are left behind to sprout. But foraging behavior is complex, and unexpected combinations of effects occur. For example, pigeon droppings left beside the plant may nourish fallen seeds, promoting their competitive advantage over seeds that were dropped without nutrient assistance. Alternatively, by dropping its still-viable seeds at a distance from the plant, the pigeon may provide a new and perhaps less competitive environment, increasing the plant's range. To one degree or another, these seed-producers and seed-consumers coevolve: The plants affect the birds, and the birds affect subsequent plant generations. Magritte painted the coevolving pigeon and the seed-producing plant decades before science, through a 1964 paper co-written by Paul Ehrlich and Peter Raven, formalized the idea of coevolution and gave it a name, giving biology a highly valued tool.[61]

By the time of the Roman Empire people had learned to exploit the foraging relationship of the pigeon and grain-producing plants, and by the seventeenth and eighteenth centuries dovecotes for housing the birds were a feature of manor life in Europe. At that point, large livestock was killed in the autumn, when fodder stores were depleted. The winter diet of salt-cured meat was supplemented by fresh rabbit, fish and dovecote residents.[62] Only the owners of the manor were allowed to have dovecotes, with their pigeonholes in which adult birds returning from the field could incubate their eggs and feed their young. It was an ingenious scheme: the lord of the manor could employ his birds as aerial collectors of an invisible tax on his tenant farmers. Once the birds filled their crops with grain protein, they returned to convert it to higher-quality meat protein in the form of squabs for the manor's dining table. The practice was eventually discontinued, probably because it became apparent that the hidden tax was extracted from neighboring estates as well as from the manor's own tenants, who were in no position to complain.

ROOM 5

Birds as a Means of Promoting Conservation

Bringing nature to the canvas is difficult when depicting organisms as structurally complicated as birds, but when it works, the effect can be profound, especially when a conservation message is conveyed. Of the images in this room, the two oldest—a painting from the cave walls of Lascaux and a coffin for a Sacred Ibis from ancient Egypt—might have been produced independent of conservation messages, but they deliver them now. Since humans are naturally attuned to notice birds in nature, artists inviting us to explore facets of avian vulnerability or the protective measures designed to offset them can appeal to bird enthusiasts and general audiences alike. When the species portrayed is overcoming heavy odds against its survival, the response can be jubilant, as it was with the "rediscovery" of the Ivory-billed Woodpecker (*Campephilus principalis*) in 2005 after a sixty-year search (see Plate 67). Hearing reports that survivors were out there was comforting, but it also made us want to see them up close.

That interest—in getting a good look at an uncommon bird, seeing its structure and plumage, and viewing an aspect of its behavior on canvas, on film, or in a photograph—has been a key element in conservation efforts. It has also sent large numbers of people out to see the living rarities for themselves. The chances of getting a good look at a bird before it flew away greatly increased with the development of the field glass. By the seventeenth century artists like Hokusai and the Dutch still life masters were recording these monoculars, the forerunners of binoculars, in art. As innovations in optics and access to remote habits accelerated, so did interest in watching wild birds. By the end of the twentieth century, Americans, for example, were spending almost as much money watching birds as watching movies and sporting events.[63]

The desire to see birds up close has also led many of us to keep them as pets within our homes and to attract wild birds to our property with food. Handouts have also kept pigeons near park benches, and ducks in the "urban wilderness" of park ponds and streams. Feeding birds has become multibillion-dollar industry, as well as a bonanza for many cats and the occasional dog. Yet the tens of millions of bird kills that suburban cats annually present to their owners and the entire bird populations on oceanic islands falling prey to feral cats have been the subjects of few paintings.

Among the celebrated images of bird-catching cats are a number of historical narratives casting them in a favorable light. In ancient Egypt, according to Herodotus, pet cats and dogs were treasured; household members shaved their heads at the death of their cat, their entire bodies at the death of the dog.[64] Cats were included in hunting scenes that convey habitat richness. Cats and birds together as an icon for habitat richness are also found in art at Pompeii, for example, and in Renaissance and baroque paintings, where domestic scenes included them among stockpiles of dead quarry in the European pantry. Images of dogs were also included in some ancient Egyptian art and in European narratives of the hunt, but unlike cats, who are natural bird predators, dogs have been artificially selected to enhance their retrieving skills, making them adept at sprinting, pointing, seizing, setting, and flushing but not actually catching birds. Unlike avian-rich landscapes, which convey beauty and attract patrons, narratives of loss and vulnerability can be difficult to market, leaving fewer artists in a position to produce them.

According to Herodotus again, both wild and tame fauna were treasured in ancient Egypt. He wrote: "There are not a great many wild animals in Egypt.... Such as there are—both wild and tame—are without exception held to be sacred. ... The various sorts have guardians appointed for them.... Anyone who deliberately kills one of these animals, is punished with death; should one be killed accidentally, the penalty is whatever the priests choose to impose; but for killing an ibis or a hawk, whether deliberately or not, the penalty is inevitably death."[65]

Occasionally the presence of a species in artwork suggests traditions of protection. Certain birds in some cultures, like cranes in India, were venerated or seen as symbols of luck; the birds were protected because of the shared belief that harming them would bring bad fortune. Occasionally the absence of a species in artwork suggests failures of protection. Artists, by creating a visual record of species known to them, have left data on species either lost from portions of their range or lost completely. Some of these extirpations and extinctions occurred with little forewarning. Others came amid efforts that were implemented too late to reverse declines or proved too difficult, too expensive, or ineffective. Art featuring threatened or extinct exotic birds is often majestic; the same handsome species that make good objects for trade or possession make good subjects for art.

But once in a while, public sentiment is aroused by a species lacking majesty, like the Dodo (*Raphus* sp.). One study concludes that since 1500 an estimated 500 bird species have been lost.[66] Among these, the Dodo is probably mentioned most often, but because there are no complete specimens, records of its appearance are not much trusted. It is not surprising, then, that the discovery of Dodo skeletons in 2006 made world news, nor will it be surprising if artists are inspired to update our impressions of the bird—to correct history while conveying a conservation message.

Occasionally a painting of a bird can become iconic and conceivably influence its fate. For example, Audubon painted the Passenger Pigeon (*Ectopistes migratorius*) and the Eskimo Curlew (*Numenius borealis*). Both were hunted in excessively large numbers, but his portrayal of the curlew had a conservation message. The Passenger Pigeon, whose population when Columbus sighted land (today's San Salvador) was an estimated 3–5 billion and constituted perhaps 25–40 percent of the avian population, was extinct by 1914, sixty-three years after Audubon's death, but the Eskimo Curlew, like the Ivory-billed Woodpecker, might still survive.[67] The practice of including exotic pet birds like macaws, parrots, and cockatoos in portraits and estate paintings increased their popularity and encouraged the pet trade—both legal and illegal —that eventually put wild populations at risk (see Plate 66).

The photographic record that documents excessive hunting or capture for trade is broad. It is certainly larger than the record of graphic images documenting these losses or losses from disease, pollution, or competition with nonnative species. Here, too, given the small market for purchasing such narratives, few artists have been able to afford to produce them.

PLATE 31. *The Shaft in Lascaux Cave* (detail), Montignac sur Vezere, France, c. 15,000–10,000 BCE

PLATE 32. *Modified Detail of the Shaft in Lascaux Cave*, 1997/2007, by Darryl Wheye

PLATES 31 AND 32 **Prey versus Prayer in Lascaux Cave**

Art

About 12,000 years before the invention of pictographic writing, Paleolithic artists in what is now France painted the only human figure in Lascaux Cave, in its deepest section, alongside a bird and a wounded bison. There is agreement that the human is male and that the bison's wound has released loops of intestine, but there has not been much discussion about the possible role of the rhinoceros (portrayed further down the wall) in the scene. And there is some disagreement about whether the man is a hunter or a shaman, about whether or not he is wearing a mask, and about why the bird is there. If the figure is a hunter with a covered head, the mask could be a disguise, allowing a closer approach to his target, perhaps the way the Kalahari Bushmen mimic ostriches when hunting them (see Plates 10 and 11). But we cannot determine if the artist is telling a story about the hunt of a particular bison that has taken place or if the artist is trying to influence a future hunt. If the latter, and if the figure is a shaman wearing a mask, any interpretation would be very speculative. We are, however, reminded that the ancient Egyptians deified the falcon because of its keen eyesight and incorporated its form into major works of art, like the Chefren and Nectanebo II statues (see Plates 4, 5, and 6). The mask resembles the head of a large raptor, such as an eagle, as shown in Plate 32, so if the figure is a shaman, then the image could convey extraordinary sight or insight.[68]

The bird is puzzling. It could be a perching bird, possibly a raven that is using the hunter as an indicator of nearby prey and hoping to snatch scraps from a potential carcass.[69] But perhaps it is the reverse: it could be a raven or a raptor whose service as an indicator of the proximity of possible edible prey was an integral part of the hunt. Or perhaps it is a Cattle Egret (*Bubulcas ibis*), which had a commensal relationship with the bison comparable to the one it has with cattle today. In this type of commensal relationship, the mammal inadvertently flushes prey that the bird can catch and the bird may inadvertently alert the mammal to approaching predators (Plate 32). Here, however, the bird could have indicated approaching predators to the bison, the hunter, or both.

Lascaux Cave has been closed to the public since 1963, but those wishing to assess the artwork to better understand the bond between hunters and birds can visit Lascaux II, a replica housed within a concrete blockhouse buried in a quarry 200 yards (180 meters) from the original site, or take a virtual tour of the real cave online.[70]

Science

Paleolithic artwork allows a good deal of interpretive leeway. Much of what we say here is open to question, but whether or not humans used birds in hunting for large mammals, it is probably safe to assume that they valued birds as important resources. Golden Eagles (*Aquila chrysaetos*), portrayed in Plate 32 to convey the idea of a bird mask, were eaten, and their bones were used for needles and needle cases, tubes, flasks, birdcalls, and flutes. The bones of other large birds were used as well—the avian species found in middens at Paleolithic sites near Lascaux include grouse, partridge, ducks, swans, bustards, cranes, storks, vultures, and eagles. Examinations of these remains suggest that our protein-needy ancestors made more complete use of the birds they hunted than we do today. Paleolithic hunters appear also to have left avian habitats relatively intact, possibly through no special effort on their part. Jared Diamond notes that New Guineans, who "depended on stone technology until European colonization began in the nineteenth century . . . are human: neither animals, nor paragons, but human. . . . [T]here is little conservation of species viewed as community property." Like the rest of us, he concludes, "New Guineans kill those animals that their technology permits them to kill."[71]

PLATE 33. *Coffin for an Ibis,* Egypt, possibly from Tuna el Gebel, 305–30 BCE

PLATE 33 **The Sacred Ibis Lost from Egypt**

Art

The ancient Egyptians mummified more than people. They wrapped up favorite pets (dogs, cats, monkeys, and gazelles) to take into the afterlife, and they offered them, along with baboons, raptors (whose mummified remains exceed a hundred thousand), Sacred Ibis (*Threskiornis aethiopicus*), and other animal species to individual gods. Pilgrims purchasing mummies as offerings probably had motives similar to those of parishioners lighting candles in a church. X-ray analysis has revealed that occasionally ancient Egyptian mummies purchased as offerings were fakes, containing only body parts or rags. Nonetheless, very large numbers of genuine mummified Sacred Ibis, apparently supplied through a captive breeding program, have been found, especially at Tuna el Gebel and Saqqara.[72]

Ancient Egyptians revered the Sacred Ibis as an incarnation of the god Thoth, who was the scribe of the gods and the creator of language. Animal cults were common throughout ancient Egyptian history, and by the Late Period and Ptolemaic (Greco-Roman) times, mummifying ibis and sealing them into conical pottery jars was a cottage industry. At Saqqara near Memphis, where a sharp bend in the Nile made it easier to control river traffic, massive catacombs were built. Within them, more than 1.5 million birds were entombed—in jar after jar, arranged in row after row in chamber upon chamber and gallery upon gallery. Votive objects, including amulets and ibis statuettes, were sacred to the cult of Thoth, and these were buried, too. Occasionally, as we see here, the mummy was housed within an elaborate ibis-shaped coffin with a removable lid on the back.[73]

Science

Why the Sacred Ibis captured the imagination of ancient Egyptians and came to be as venerated is not obvious. But there is room to speculate based on some scientific knowledge gained about the birds since the ibis here was produced. Perhaps the Egyptians had a practical interest in the ibis's ability to kill snakes or to rid their shops of vermin, or perhaps they had an awareness of the importance of its habitat. (Artists often depicted the bird among clusters of papyrus flowers—papyrus was used to make the paper of the era and was considered a significant resource for Thoth.) Possibly the ibis starred in a now-lost legend, like the one attributed to Pliny the Elder (23–79), who claimed that the bird used its bill to self-administer a water enema. Pliny may appear credulous in this instance, but the bird does perform a similar behavior when collecting preening oil from the gland near the base of the tail and distributing it. He told other tall stories about nature—even though he was a stern and sometimes unpleasant critic of similar questionable stories that had been reported by Aristotle. Still, there is some suggestive evidence for the enema formulation. In 1873 the Egyptologist George Ebers (1837–1898) purchased a papyrus scroll from a dealer who claimed to have found it between the legs of a mummy. The text, known now as the Papyrus Ebers, consists of 108 columns divided into forty-five groups. The second group includes prescriptions, some of which were to be taken as enemas. It is unclear how enemas were administered, but a horn with a hole in the tip is housed in the Louvre. According to Herodotus, the Egyptians assumed that all disease was foodborne, so for three successive days each month they purged themselves using emetics and enemas (clysters) to administer drugs.

In any case, the Sacred Ibis, once venerated, was relatively common throughout Egypt until 1800, but it was lost from the country during the 1870s and is now limited to areas south of the Sahara and the northern end of the Persian Gulf. That humans are unable to successfully re-create the captive breeding program for this bird and successfully reintroduce it suggests the degree to which its natural habitat has been modified.[74]

PLATE 34. *The Peasant and the Nest Robber*, 1568, by Pieter Brueghel the Elder

PLATE 34 A Conservation Message from the 1500s

Art

In Brueghel's day, the densely populated port city of Antwerp was the center of international trade. It attracted ambitious capitalists—who, in turn, supported the city's artist guild and its 300 active members. Surplus capital was also shared with Charles V (through high taxes) and invested in new industries in the provinces, which made the rich far richer. It also made the workers far more vulnerable because jobs exported to small provincial towns and the countryside were not shielded by the guilds that traditionally protected worker wages and working conditions. It was a vibrant era, but the capital-driven changes in the production and distribution of goods and thus in labor led to conflicts that in a number of ways resemble those we see between capital and labor today. Pieter Brueghel the Elder recorded much of this on canvas.[75]

His work often has a strong moral tone; he was especially critical of pilfering and of those who turn a blind eye to it. This scene—of a peasant pointing behind him toward a nest and a nest robber—includes symbols that seem intended to remind viewers of his day to think twice before stealing eggs or tolerating those who do. On the right, a malformed willow is said to represent the nest robber and identifies him as a misfit overcome by impulsive greed. On the left, the iris and its blade-like leaves (a model for the heraldic fleur-de-lis) apparently stood for strength gained through taking responsibility, and the bramble behind it stood for power achieved in overcoming temptation. Of Brueghel's paintings featuring birds, this one has an especially explicit conservation message.

Science

Unfortunately, Brueghel's message still needs to be heeded. Pilfering continues despite laws safeguarding nesting birds, especially endangered ones. The nests of some rare birds have required round-the-clock surveillance to protect them from human hands despite the threat of stiff penalties. The Osprey (*Pandion haliaetus*), for example, a summer visitor to the United Kingdom, became extinct there at the end of the nineteenth century largely because of egg collecting and trophy hunting. When a pair returned to breed in 1954, the Royal Society for the Protection of Birds (RSPB) successfully provided surveillance of the nest site that kept watchers engaged for twenty-four hours a day every day of the week.[76] After half a century the number has grown to 200 breeding pairs.

PLATES 35 and 35a. *Untitled* [The banding of a heron] and detail, probably 1764, by an unknown artist

PLATES 35 AND 35A **Banding Birds in the 1700s**

Art

The brass band in the hat depicted here reads "1764," and it will encircle the leg of the Gray Heron (*Ardea cinerea*), the European form of North America's Great Blue Heron (*A. herodias*). Although this painting was produced nearly two centuries after the oldest metal banding recorded—an event that occurred in France, around 1595, when one of Henry IV's Peregrine Falcons was recovered 1,350 miles (2,170 kilometers) away from where it had been banded—it predates widespread banding efforts, and the novelty of its subject matter may have generated considerable interest. Today it generates interest and some mystery. It serves as the frontispiece for *Bird Trapping and Bird Banding*, by Hans Bub, but efforts to identify the artist and the current copyright holder have failed.

The modern banding of birds dates back to the success in 1710, when a Gray Heron bearing several bands—one from Turkey, more than 1,200 miles (1,900 kilometers) away—was recovered by a falconer in Germany. Eighteen years later, in 1728, the band that Duke Ferdinand had wrapped around the leg of a bird back in 1669 or so was recovered. In North America the first known banding occurred in 1803. The next year John James Audubon found two phoebes near Philadelphia bearing the silver cord he had tied to their legs the previous year, when they were still in their nest. (The banded nestling phoebes—called Peewee Flycatchers in Audubon's day—became special friends of Audubon's; he frequently visited them while courting Lucy Bakewell, later his wife.) These isolated incidents were among the first faltering steps toward using a monitoring technique that took a century to catch on but eventually became a worldwide activity. Today thousands of serious students of bird biology participate in banding, and it contributes substantially to our knowledge about birds.[77]

Science

In Europe the development of systematic bird banding is attributed to a Dane, Hans Mortensen, who began with European Teal (*Anas crecca*), Pintail (*A. acuta*), White Storks (*Ciconia ciconia*), Starlings (*Sturnus vulgaris*), and various hawks in 1899. Back then, there was no guarantee that the bird bands, if found, would be sent back to Mortensen. Within three years, in 1902, Paul Bartsch had instituted systematic banding in North America using bands etched with the words "Return to Smithsonian Institution." Soon after that, in 1909, the American Bird Banding Association was organized. In 1920, U.S. and Canadian governmental agencies took over the program, and now bird banding is a coordinated global effort that provides data on bird behavior, survival, and travel.[78] Banding serves other purposes, too. For example, it allows the identification of individuals in field studies where researcher interference must be minimized. (In cases where interference must be reduced even more, microphones are used to record voice prints that can be analyzed electronically to identify species and, in some cases, region of origin.)

As interest in migration and dispersal have intensified, scientists have banded birds to track them as they travel to find resources or mates, as they move to their breeding grounds, or as they just leave home when habitat and climate change threaten their survival. To find out where birds have been, researchers can analyze genetic differences among birds, chemical signatures of feathers, and routes taken by birds wearing minute tags keyed to global positioning systems (GPS). These sorts of analysis are enriching the field of movement ecology, which is building a framework to help us assess how birds navigate and how changes in patterns might reflect changes in environmental quality.[79]

PLATE 36. *Mossy Branches—Spotted Owl*, 1989, by Robert Bateman

PLATE 36 An Influential Wildlife Painter and an Avian Icon

Art

In the forest the piercing whine of chain saws has stopped, and the thunderous vibration of falling tree trunks has been stilled. Intense shafts of sunlight now slice through the tempering veil of the canopy, catching only peaceful motes of forest dust rather than the thick eddies found during logging days. The stare of the owl in this painting challenges us: "Now what?"

In the 1980s environmentalists catalyzed public debate about the future of the Spotted Owl (*Strix occidentalis*), the future of its old-growth forest habitat, the future of the loggers working there, and conditions that might allow all three—birds, forest, and loggers—to coexist. It was not easy to convey the vulnerability of the bird and the visceral feel of the ancient old-growth forest to people who ordinarily walk on sidewalks, and it was equally difficult to persuade loggers that the forest ecosystem could no longer absorb their current methods of extraction. Teasing apart the arguments that were keeping both sides at loggerheads was a delicate task, but spokespersons who could deliver authoritative information and artists who could deliver compelling images stepped up.

Robert Bateman, with 2,000 images featuring North American animals and their habitats to his credit, has been markedly successful at converting environmental issues into art. "For me," he says, "the preservation and celebration of the natural world is the underlying motive for virtually all of my art." The notion that saving the Spotted Owl saves the forest and the 350 bird and mammal species found there was obvious to Bateman, but it would take more than a beautiful bird portrait to make his point. That is why his owl is out on a limb. As Bateman notes, "I wanted [it] to be in your face."[80]

Science

The Spotted Owl is a finicky bird. It cannot breed every year, it cannot tolerate heat, and with a home range of 1,400–4,500 acres (550–1,800 hectares) it cannot survive in small patches of habitat. Very few of its young live long enough to breed, and those who do compete poorly with other owls, such as the closely related Barred Owl (*Strix varia*). Studies on Spotted Owl behavior and resource requirements by U.S. Fish and Wildlife Service biologists leave little doubt that logging the North American temperate rain forests until it is broken into small plots will lead to the owls' extinction there. Even before the last owl is lost, however, its functional role within its community will have petered out.[81]

Mezzanine
Thinking about Aesthetics, the Oldest Bird Paintings, and Painting Nature

The linkage of science and art has historical as well as contemporary origins in the pictorial archive, which raises a question: Does linking the two cast doubt on the often-cited "Two Cultures" theme of C. P. Snow that modern scientists and artists are members of opposing cultures that do not exchange much information?[1] David Edwards, in *Artscience* (Harvard University Press, 2008), finds the traditional line between these two cultures still firmly drawn and argues eloquently for a new intellectual milieu that will be an interdisciplinary catalyst for innovation and creativity. Others find the boundary more porous. For example, in 1995 the physiologist Robert Root-Bernstein and colleagues asked thirty-eight scientists about art, science, and the Two Cultures premise. The results showed an interesting split. Influential and recognized scientists (eleven National Academy of Science members and four Nobel Laureates) discounted the Two Cultures divide, citing as examples a number of productive scientists who were also involved in the arts. Scientists toward the other end of the influence spectrum, however, were likely to believe that artistic and literary avocations would provide distractions that could impede careers.[2] If these thirty-eight responses lead to any conclusion, it is that (1) any estrangement is likely to be a matter of a time constraint and (2) the barrier separating art and science fluctuates and develops leaks. So why is conventional wisdom apt to characterize science and art as isolated, polar opposites? It may be because conventional wisdom is focused on process—the actions involved in doing science or art—rather than on the fundamentals involved in both activities.

The processes *are* decidedly different. Although both the scientist and the artist usually start off by framing a question or proposing a plan, their actions quickly diverge. Most scientists proceed by formulating hypotheses and testing them, taking steps to ensure the possibility of rerunning experiments and replicating the results. Most artists intend to create works that possess a unique appeal and are therefore one of a kind. The scientist who intentionally replicates the scientific experiment of another receives recognition for either validating or invalidating the results of the previous work. In contrast, an artist's unacknowledged intentional replication of a work of art is dismissed as either a useful training copy (see Plates 19 and 20) or a forgery.[3]

Nevertheless, the fundamentals underlying science and art do have important domains of commonality that fuse the two in interesting ways. Prominent among these commonalities are inspiration, creative curiosity, imagination, and the unconscious. All of these provide for flow through the permeable Two Cultures barrier. Contributing to the flow are those individuals who do both science and art or who are interested in both.

Thinking about Aesthetics

Repeatedly, those who have made their mark in science describe the work as part imagination, part logic. The mathematician Henri Poincaré (1854–1912) wrote that among scientists, invention is driven by intuition and proved using logic. A century later, Peter Medawar (1915–1988), who won the Nobel Prize in Medicine, provided almost exactly the same summation in the book he intended to be read by beginning scientists. In his view, even though scientists start out with a transitory collection of tentative theories based on hypotheses and then rely on common sense and experimental skill to test those hypotheses and build theories to explain how the natural world works, it takes imagination and creativity to formulate the hypotheses in the first place.[4] Two physicists, Max Planck (1858–1947) and Albert Einstein (1879–1955), weighed in on the side of the imagination as well, with Planck declaring that a groundbreaking scientist needs a vivid, intuitive, artistically creative imagination in addition to an eye for details. Einstein went a little further: "The mere formulation of a problem is far more essential than its solution, which may be merely a matter of mathematical or experimental skill. To raise new questions, new possibilities, to regard old problems from a new angle, requires creative imagination and marks real advances in science."[5]

Of course, artists make use of these qualities. John Constable (1776–1837) is considered the first artist in England who thought landscapes ought to be based on observable facts, what he called "natural painture," but he conceded that imagination still had a significant role to play: "The whole object and difficulty of the art (indeed of all the fine arts) is to unite imagination with nature." He was careful to point out, however, that imagination alone would not suffice: "I hope to show that ours is a regularly taught profession; that it is *scientific* as well as *poetic;* that imagination alone never did, and never can, produce works that are to stand by a comparison with *realities*."[6] A century later, Paul Klee proposed that "art does not reproduce the visible; rather, it makes visible" (see Plates 41 and 42).[7] Art that portrays nature in a scientifically valid way benefits from the views presented by both of these artists. It maintains its faithfulness to biological reality but permits major imaginative departures in other aspects of the work.

Reliance on the unconscious has a long history. When the source of an inspired thought is unclear, some people conclude that the idea came "in a flash," as if it had been generated externally. An interpretation of that kind would have disappointed the Roman philosopher Plotinus (205?–270), who believed that "the absence of a conscious perception is no proof of the absence of mental activity."[8] He reminds us that none of our ideas are free from unconscious influences; none are neutral. Scientists, after experiencing a flash of insight, are likely to try to retain as much neutrality as possible while they test the hypotheses against new experiments and then interpret the results. Artists, in contrast, can proceed in a relatively unconstrained way because they have neither need nor incentive to filter out their emotions. In fact, some doubt that it is even possible to filter out conscious emotional content. Leonard Meyer, who writes about musical meaning and communication, takes the view that emotion and intellect are inseparable and that neutrality is impossible. Arthur Koestler (1905–1983) put it this way: "The scientist's discoveries impose his own order on chaos, as the composer or painter imposes his; an order that always refers to limited aspects of reality, and is based on the observer's frame of reference, which differs from period to period as a Rembrandt nude differs from a nude by Manet."[9]

There are times when artists invite us to inspect nature through a scientific lens and see an otherwise latent truth that has been revealed by researchers. They spare us the effort of reading terminology-laden reports but inevitably add a patina to the narrative they depict. That patina varies depending on their intent, their medium, their style, and the depth and breadth of their perception. But no matter how successfully artists make their chosen slice of nature visible, they must first catch the viewer's eye and then hold it. After 30,000 years, we are still not quite sure how this is done. There are, however, some clues.

When artists capture our attention and draw us in, they seem to take one or more of three approaches. They may engage us by presenting an essential aesthetic quality—beauty, elegance, novelty, harmony, order, or pattern—either alone or in combination. They may present a titillating image, or part of one—a gaping wound, an act of seduction, a pending act of violence, an aspect of nurturing. Finally, they may catch us through the construction of an improbability —like the Mona Lisa's seemingly shifting smile (see Room 7)—or an out-and-out impossibility, like M. C. Escher's going-nowhere staircase (see Plate 45). However those aesthetic or arresting qualities catch and hold our eye, the successful artist manages to lure us —objectively, subjectively, or both—with a particular signal or feature or with several signals or features that pull us in simultaneously and keep us looking as we develop multiple interpretations of what we see.[10]

The individual aesthetic qualities themselves —the features that attract us—are not clearly understood. This is not owing to lack of effort. In the days when biology was developing into a systematic science, some 250 years ago, the French encyclopedist Denis Diderot (1713–1784) addressed the question of aesthetics. At the time, the importance of nature was reemerging in the West. It had been suppressed in medieval thought, where it was assumed, like all matter, to be separate from God and therefore of relatively little importance, but began working its way back into prominence during the Renaissance. Diderot reasoned (it was the Age of Reason, after all) that nature was the only truth and that the responsibility for aesthetically conveying its vitality and constant fluctuations should fall to artists. He also reasoned that the perception of beauty was neither innate (uncultivated taste) and independent of reason, nor a simple awareness of sensations. Instead, he saw it as a complex thought process that was immediate (emotional) and arose from a preexisting point of view (rational).[11] His analysis retains some appeal these many years later. A variable mix of emotional and rational responses to beauty could account for the complexity of aesthetic experience, our inability to pin down the flux of the universal and the culturally based qualities that define it, and its plasticity—that is, its relative constancy but its coincident ability to change.[12]

There is a boundary between a work of art and its viewer. Dissolving it reduces our resistance to entering the work and viewing it from within the scene or identifying with the relationship depicted. As we will see in the Upper Gallery, this allows a momentary departure from our position as a bystander (see Plate 46). If the painting is convincing, we might take away a refined understanding of an important relationship in nature—and might even shed some long-held assumptions about the inherent character of our personal relationship to the natural world. In short, both scientists and artists must get others to look, to make connections, and hopefully to remember. A picture worth a thousand words is easier to remember than the thousand words it represents.

Thinking about the Oldest Paintings of Birds

When it comes to technical skill and style, what can we conclude about the oldest wall art? To answer this question the art historian Sir Ernst Gombrich (1909–2001) wrote:

> We now know that more than 30,000 years ago ice age artists had acquired a complete mastery of their technical means, presumably based on a tradition extending much further into the past. This tradition had equipped them with serviceable conventions in the rendering of various species, but it had not prevented them from branching out on their own—witness the unique portrayal of an owl (or of auks at Cosquer), not to mention various fantastic creatures, one of which looks like a minotaur—a bull with two human legs—possibly representing a masked shaman. In any case, these early hunters must have felt free to experiment with frontal views, rudimentary foreshortening, and the device of shading to enhance the impression of rounding forms, if not, perhaps, of the fall of light. The vocabulary they handled with such supreme artistry can now be seen to have lived on in the formulas, not to say stereotypes, painted or scratched on the walls of such caves millennium after millennium. [See Plates 2 and 18.][12]

Gombrich's description of technical mastery and his observation that these animal images have served as models for future artists set the stage for thinking about image content that few will fault. When assessing the content of the images, however, it is more difficult to reach a consensus. General agreement about whether the artists were recording animals encountered on a past hunt, depicting them in the hope of encountering them on a future one, or creating a random, serendipitous rendering based on the contours of the rock face is lacking. Like other examples of Science Art, we interpret cave animal art as all three, sharing the view of cave art authorities like the historian William McNeill, who suggests that rather than thinking in terms of either-or propositions, it might help to think of the art as a mix, partly as a record of real ecological relationships, partly as a portrayal of animism (the belief in invisible spirits), and partly as a reading of the wall face. McNeill wrote: "Cave art derives both from the natural world of flesh, blood, and brain that once existed on the Mammoth Steppe, and from an imaginary world of invisible animal spirits, embodied and disembodied, who, the artists believed, controlled, directed, and inspired animal and human behavior both above and below ground. Only by positing such an imaginary world can we begin to understand the paradoxical mix of serene and accurate masterworks with the multitude of free and spontaneous scribbles that together comprise the art in the caves." He also noted: "Everyone agrees that many of the masterworks of cave art were constructed around preexisting marks and curves on natural surfaces. In that sense, we can say that human intervention . . . assisted them to emerge from the stone."[13] Keeping this multipurpose perspective in mind can help when viewing the birds in Paleolithic cave images—the owl in Chauvet Cave (see Plate 2), the owl family in Les Trois Frères Cave (see Plate 24), the auk in Cosquer Cave (see Plate 17), and the perched bird and probable bird mask in Lascaux (see Plate 31).

Thinking about Painting Nature

Why can't painters just sit down and paint nature, recording what they see the best they can? We return again to Gombrich, who wrote:

The simple demand that they should "paint what they see" is self-contradictory. . . . We have often looked back to the Egyptians and their method of representing in a picture all they knew rather than all they saw. Greek and Roman art breathed life into these schematic forms; medieval art used them in turn for telling the sacred story, Chinese art for contemplation. Neither was urging the artist to "paint what he saw." This idea dawned only during the age of the Renaissance. At first all seemed to go well. Scientific perspective, *"sfumato,"* Venetian colours, movement and expression, were added to the artist's means of representing the world around him; but every generation discovered that there were still unsuspected "pockets of resistance," strongholds of conventions which made artists apply forms they had learned rather than paint what they really saw. The nineteenth-century rebels proposed to make a clean sweep of all these conventions; one after another was tackled, till the Impressionists proclaimed that their methods allowed them to render on the canvas the act of vision with "scientific accuracy."

The paintings that resulted from this theory were very fascinating works of art, but this should not blind us to the fact that the idea on which they were based was only half true. We have come to realize more and more, since those days, that we can never neatly separate what we see from what we know. . . . In fact, as soon as we start to take a pencil and draw, the whole idea of surrendering passively to what is called our sense impressions becomes really an absurdity. If we look out of the window we can see the view in a thousand different ways. Which of them is our sense impression? But we must choose; we must start somewhere; we must build up some picture of the house across the road and of the trees in front of it. Do what we may, we shall always have to make a beginning with something like "conventional" lines or forms. The "Egyptian" in us can be suppressed, but he can never be quite defeated."[14]

We agree: The artist's style, what the artist saw, and what the artist might have known beforehand all influence what the art records. We would add, however, that when assessing influences, viewers also consider what the artist might have been allowed to express.

A further degree of uncertainty is introduced when looking at examples of Science Art produced by an artist whose style moves away from realism, which can call into question the artist's accuracy. If we agree that a scene or an event outside the window can be seen in many different ways and recorded in many different styles, the question of accuracy arises only when it prevents viewers from registering a truth about nature. Viewers can significantly reduce their uncertainty when that truth is noted in the caption.

The images in the Upper Gallery have been selected to further the discussion of Science Art as an underrecognized category of art (Room 6). The focus is on the issues of content, style, and medium (Room 7), captions (Room 8), and venues (Room 9).

Upper Gallery
How Science and Art Overlap

Both scientists and artists try to persuade others to make connections and remember them. To help viewers come away with more than a vague sense of "maybe I understand it," artists whose works fit the criteria of Science Art eliminate the "work it out yourself" directive implied by many untitled paintings. Although they may have created the works for their own satisfaction, the images portraying birds must be neither so cryptic nor so personal that viewers cannot decode the biology that underlies them. (The subject matter is limited here to birds for the sake of our discussion.) These artists can reduce our uncomfortable sense of not knowing how to interpret a picture, and relieve us of doubts about whether we have been sufficiently attentive to take details into account. A viewer can always dig deeper into the message and attributes of an image, but a well-constructed caption makes it possible for the viewer to avoid walking away confused.

Sometimes a vague feeling of disappointment settles over us when we look at art. But with Science Art we can look forward to finding a degree of engagement. Walking away from a painting that we feel we understand is not unlike the satisfaction of putting down a well-written mystery novel confident of who did what, how, and why.

This kind of art tells a story, but both recognizing the science and getting involved in the artistic dimension may require separate readings. In the case of bird art, even the ornithologically naive can make observations and draw conclusions because there is a lot that a viewer can figure out without consulting the caption. Gauging and internalizing the scope of the narrative just takes a second. We can instantly note the level of organization (Does the subject represent an individual, a pair, a family, a group, a population, a community, or a specific ecosystem?) and determine what appears to be happening. Whether or not we recognize the bird, we can note its phenotype (its plumage and coloration, bill shape and size, leg length and foot webbing, wing length and shape, and so forth), which allows us to deduce whether the bird needs to remain cryptic to predators, boldly attractive to mates, neither, or both; whether it uses its bill to tear flesh, snap up insects, crush seeds, snare fish, pluck fruit, or sip nectar; whether it wades, swims, or spends little time in water; and possibly whether it soars, glides, or mostly flaps its wings while airborne. Since the images are decipherable, the artistically naive are well positioned to assess the art—the compositional elements, the patterns, and the use of space, as well as the artist's techniques, and so forth, and make determinations about the aesthetic power of the piece, its emotional appeal, and its ability to engage. Once we have a sense of the bird and the narrative and a feel for the artist's style and use of medium, the caption can supply more information. Because captions play such a central role in Science Art, they are the focus of Room 8, with additional material provided in Appendix 2.

The requirements of Science Art are simple: It must remain faithful to the known scientific facts, portray them artfully, and give viewers a chance to understand its significance. It should also provide a caption that will set viewers straight if they do not understand it, a caption that also helps viewers fit the image into the great clamor of creative work *and* relate to it in a wholly personal way. The aim of the Upper Gallery is to make it easier to recognize works of Science Art and to benefit from the science they contain.

Room 6 has ten images that can help us understand why these renderings seem to work as they do and how the blending of science and art to portray nature can enhance both without diminishing either. Several images will show how the work of polymaths skilled in both science and art and how the collaborative efforts between scientists and artists can produce highly instructive art. The remaining images provide a platform for exploring the role of aesthetics in Science Art and the old schism between illustration and art.

Room 7 has ten images that suggest how the content of an image, the artist's style, and the chosen medium influence Science Art. For example, to decipher the narrative we will need to identify the particular behaviors portrayed. If the bird is thermoregulating—trying to lose or retain heat—that could explain why it is standing on one leg. If it appears to be sleeping, it could be sound asleep or just dozing while it keeps an eye out for predators or guards its mate to protect against cuckoldry.[1] We will also need to anticipate the probable outcome of interactions that might possibly explain the motivational state of a subject whose face cannot express motivation. And we will want to appraise the habitat. After we ask these questions, the story—or several stories—presented in the image should have emerged. If not, we can check the caption. We might also assess how the artist's style and choice of medium enhance the narrative and the underlying science it reveals.

Room 8 has three images that demonstrate the importance of captions in portrayals of biological truths that might not be well known or intuitively obvious. One of these images shows how captions can conveniently correct errors—in the science or in the rendering of the science, or in the viewer's initial perception of the evident science. All three images show how in Science Art an agile mix of aesthetics and information locks in our gaze and stimulates questions that, if unanswered after close inspection, invite us to consult the accompanying caption.

When viewing these images, we should instantly be able to tell whether we are looking at more than portraits of animals and more than scenery. If it is difficult to put a finger on what else is going on, the caption should tell us who produced it and when (providing some clues), what creatures are portrayed (which may jog some memories), and what messages about natural history are being communicated (often surprising).

Room 9 has two images that are meant to hang in public hallways. The appearance of Science Art in public spaces can occur as unexpectedly as an interesting bird at an inopportune moment—and we may decide to move on without allowing our interest to be more than transient. Yet sometimes we pause. Looking again at the art years later may evoke our original response—or a very different one. The messages change as we change and as our culture changes. Art is, in short, one of our perpetually renewing cultural resources.[2] Unfortunately, art of this kind can prove discouragingly difficult to track down when we set out to find it. These difficulties can be reduced, and encouraging their reduction is one of our purposes in writing this book.

Room 10, the last room, has four images that recapitulate the art in all the preceding rooms. They remind us about the popularity of birds and the practicalities of attaching a name to this genre of art. They remind us how our innate sensory abilities create our perception of nature, how optics and photography enable artists to portray details augmenting that perception, and how perception can be heightened when it is coupled with an understanding of nature's inner workings—both in the world around us and as it is recorded by artists.

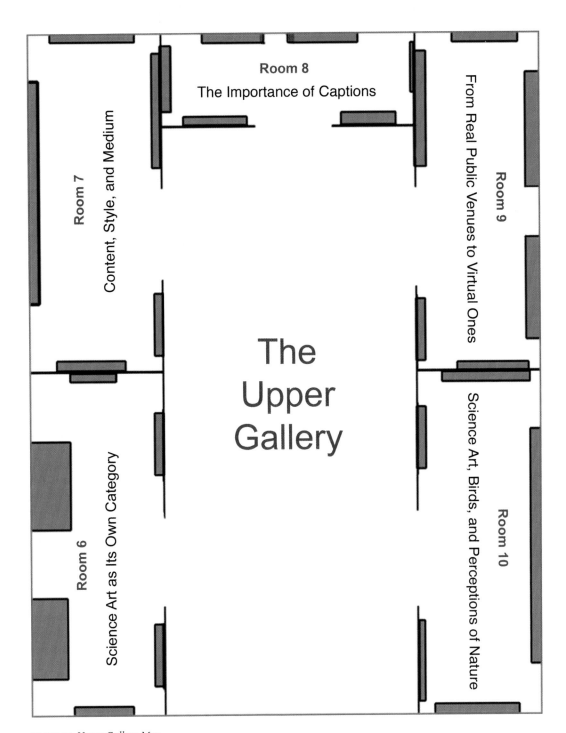

PLATE 37. *Upper Gallery Map*
The five rooms in this gallery contain 29 original images, arranged to explore how science and art blend. Rooms 6 through 9 extend the discussion of Science Art as a genre, discuss the key role of captions, and note the expanding opportunities to see examples of this kind of art, especially in public spaces and online. Room 10 serves as a summary.

ROOM 6

Science Art as Its Own Category

The images in this room show some of the advantages of recognizing Science Art as a unique category. They offer narratives about birds in nature and narratives about birds and technology; they exemplify the value of collaborative efforts between scientists and artists; they demonstrate the use of visual metaphors grounded in science; and they show the arbitrariness of the boundary between illustration and art.

As we saw in the Lower Gallery, people have been creating images that depict particular sets of relationships for at least 30,000 years, often relationships with and in the natural world. Since Paleolithic times, artists have portrayed birds and other animals as objects of aesthetic, cultural, or religious interest, or as prey, predators, possible sources of trouble, or potential companions. As Paul Shepard has written, "How we see events or issues always has a basis in the senses. Ecological reality is the arbiter of the senses and all of the impulses in brains. It is the generator of mind and the sinew of culture. It is food habits and food webs, the passage of energy."[3]

In addition to representing ecological reality, birds represent particular human values; age-old examples are the bluebird of happiness, the dove of peace, and the owl of Athena/Minerva, the Greek/Roman goddess of wisdom. Birds also represent critical resources required for human survival—resources harvested sustainably, as with the Inuit utilization of breeding ducks in the Yukon-Kuskoquim Delta in Alaska, or exploited unwisely, as with the collapse of the Easter Island ecological system. There, most land birds disappeared, and as resources dwindled, the richest seabird breeding site in Polynesia was transformed into a silent gallery for giant sculptures and hundreds of "birdman" petroglyphs (see Plates 38 and 39).

Historical depictions often reveal the artists' keen if primitive understanding of ecological relationships. Contemporary artists, in contrast, have an opportunity to take advantage of the fruits of formal research as it has broadened our understanding of behavior and ecology. Overall, the purpose of the art may be decorative—but often, too, the representation is scientifically accurate, giving us information about what kind of plant or animal is portrayed, how the plant or animal fits into the place where it is found, or what relationship it has to the human community. These images can convey a deeper message about the character and perhaps the fragility of these relationships. Science Art is about nature, and it has always been scientific as well in the sense that it records information attained through study and conveys general truths about the physical world.

Unfortunately, however, art featuring nature that meets these criteria has not formally acquired a name. We are proposing "Science Art," but searching for such art still remains hit or mostly miss. Science Art is well represented in publications, but readers will not find a master index to browse. Long ago it worked its way onto museum walls, but visitors will not find a category listed in collection inventories. It appears in galleries featuring wildlife art and in public spaces and other venues, but viewers lacking a background in biology or familiarity with the species portrayed may not recognize the scientific significance of the narrative—

and often cannot count on finding a caption to help them interpret the scientific messages. And even though examples of this kind of art are increasingly available for viewing online, Internet surfers encounter massive search results encumbered with irrelevant listings.

Not only will the adoption of a specific name make this kind of art easier to find, it will also mean that we can expect to find faithful representations of species and subject matter presented in ecologically relevant combinations of vegetation, prey and predator species, topography, climate, and so forth. Although some might balk at what they perceive to be the marriage of a rule-driven objective enterprise with an unencumbered subjective creative process, the adjective *Science* in *Science Art* allows viewers to make positive assumptions about the validity of the portrayal.

Whatever name makes its way into the vernacular, the images in this book document ways we humans have brought animals into "our" realm, often by making them into cultural or religious icons with scientifically realistic underpinnings. Either the art itself or the caption provided with it presents a basis for decoding the science and allows viewers to become more fully involved in the slice of nature portrayed or alluded to—and perhaps even to reconsider traditional lore (see Plates 27 and 28).

The stories conveyed through literature and art may provide a backdrop against which to measure our behavior. Some stories may accompany us through life; others may capture our attention only once and only briefly, when events warrant it. As the Nobel Laureate Ben Okri puts it, "Stories are the secret reservoir of values: change the stories individuals or nations live by and tell themselves, and you change the individuals and nations."[4] When looking at examples of art about nature that have science as one orienting component, we need to train our eye to look for the science. When we are unfamiliar with the science, we miss the relationships portrayed; that is, we miss a story. This is a topic of discussion in Room 9, where public space venues are covered. It is revisited in Room 10 in connection with Plate 67, which depicts the Ivory-billed Woodpecker, the subject of many articles and many works of art since its possible rediscovery in 2005. That discussion broadens our understanding of what is at risk as humans use planetary resources and how much responsibility for that use we should assume as individuals and as members of a particular society. Key to the success of employing Science Art as a teaching tool—and using it as a conduit through the barrier that tends to isolate science from art—will be the number and variety of participating artists, the number and variety of the works produced, the number and variety of venues at which the works can be seen, and the ease with which we can find works of Science Art.

PLATE 38. *Sooty Tern and Birdman Petroglyphs on Boulder Panel 64, Mata Ngarau, Orongo, Easter Island,* 2005/2007, by Darryl Wheye

PLATE 39. *Birdman Petroglyphs on Boulder Panel 64, Mata Ngarau, Orongo, Easter Island,* 1982, photograph by Georgia Lee

PLATES 38 AND 39 **Telling a Credible Story about Birds and Nature**

The Narrative

Plate 38 shows three of the 481 petroglyphs of birdman chimeras on Easter Island and a Sooty Tern (*Sterna fuscata*), one of the island's few remaining bird species. The association has meaning to those familiar with the cult that once held sway there. The chimeras relate to an annual event whose outcome affected the entire island. For hundreds of years the islanders held an extreme egg hunt. The timing of the contest each spring was determined by the stars, and the contenders included a number of politically powerful men and the young man each chose to act as his proxy in the competition. To win, a contender's proxy had to be the first to retrieve an egg from a Sooty Tern colony on Motu Nui, an offshore islet. When the proxy handed over the egg, the waiting contender became the new birdman and spent the rest of the year sequestered. Throughout the year the birdman was in a position to make significant decisions. He was authorized, for example, to select several individuals to serve as sacrifices, an act ensuring future acts of vengeance and subsequent rounds of hostilities.[5]

Why proxies were chosen for the expedition to Motu Nui quickly becomes clear. First they descended 1,000 nearly vertical feet (300 kilometers) from the rim of a million-year-old volcano to the beach below. Then, with the aid of bundled reeds that served as floats, they swam a mile (1.6 kilometers), toting their provisions through strong, swirling, shark-infested currents. The proxies who reached the islet sheltered in caves and awaited the start of egg laying and their chance to retrieve one of the well-camouflaged tern eggs and take it back to Easter Island.[6]

There is some debate over the dates associated with the rise and fall of the birdman cult (c. 1200 to 1860) and over details related to the rituals, but there seems to be a consensus that the annual birdman ritual "reconfirmed the social order, reinforced status and power, and became a rallying point for the society."[7]

Viewing the Science

These petroglyphs reflect real relationships in nature and real consequences of the mismanagement of resources. The lesson of Easter Island resonates with an increasingly large audience, in part because of the work of dedicated anthropologists and because of discussions like that of Jared Diamond's in *Collapse* and in part because bird study and bird watching have become an exploding avocation for people in all walks of life in North America, Europe, and many other parts of the world. As Francis X. Clines noted in a *New York Times* article in 2001, "the more than 50 million birders in the United States (some 60 million, according to the National Audubon Society) spend over $25 billion each year on feed, equipment, and bird-related travel.[8] As more artists produce narratives that place a high value on scientific accuracy—by providing credible representations of their subjects' locale, behavioral features, and ecological relationships—new generations can increasingly learn from the past.

PLATE 40. *Honeybees, Honeyguides, and Honey Hunters*, 2004/2007, by Darryl Wheye

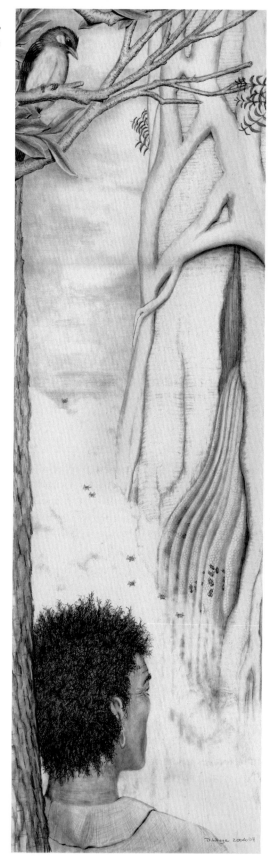

PLATE 40 **Telling a Complicated Story about Birds and Nature**

The Narrative

This image portrays one of the oldest human-avian relationships we know of: between the Boran people and the African Greater Honeyguide (*Indicator indicator*). It shows a beehive built into the trunk of a tree encircled by a Strangler Fig. A man is waiting for a few bees outside the hive to depart or become inactive and for the smoke from his torch to filter further into the hive, where it will inactivate the bees inside tending eggs and larvae. The passivity of bees in response to smoke seems paradoxical, but it possibly occurs because smoke masks alarm pheromones and because bees might have evolved a "shelter-in-place" response to forest fire smoke.[9] As soon as the bees are calm, the man will harvest their honey. The perched honeyguide, who led the man to the hive, is, like the man, watching and awaiting its chance to feast on the beeswax.

Viewing the Science

A caption decoding the science is essential for many viewers. A comprehensive caption might read as follows: "The hives of African honeybees provide a rich resource for wax-seeking Greater Honeyguides and honey-seeking people. In northern Kenya the birds (who are adept at locating hives) and the nomadic Boran people (who are less so) have evolved a symbiotic hunting relationship. The honeyguide reduces the hunting time of the Boran by conveying hive distance and direction through calls and through changes in perch height and flight pattern while heading toward the hive. This assistance provides its own reward. After the honey hunters crack open the hive and drain its honey, the honeyguide forages on the broken honeycomb, which would have been otherwise inaccessible.

"Does this relationship qualify as coevolution? The evidence points that way. Rock art produced thousands of years ago shows humans robbing hives for honey, and the first written records of avian assistance were made in the 1600s.[10] A three-year study found that the Boron hunters took, on average, about nine hours to find hives on their own and about three when guided by a bird; 96 percent of the nests were accessible to the bird only after the hunters opened them. The study clearly suggests that the interaction is purposeful on both sides.

"Apparently the duration of the bird's disappearances as it flies from perch to perch, the distance between its perches, and the perch height that it selects while awaiting the hunter all decrease the closer the bird gets to the hive.

"Are the birds and hunters cooperating these days? Less and less: the Boron people are shifting away from honey to commercial food, sugar, drugs, and alcohol, and the birds are increasingly restricted to large nature reserves that exclude the Boron and other subsistence hunter-gatherers. Coevolution in this case is quite likely cultural. In contrast, if coevolution between these birds and other honey-seeking mammals, like Africa's Honey Badgers, exists, it is probably genetic."

Many artists are attentive to routing the viewer through the pictorial narrative, just as many writers are attentive to routing the reader through a plot. Here the artist tried to route viewers in an elliptical pattern to help them avoid missing important elements of the action involving the man, the bees, and the smoke, but at some point viewers will need to break away or miss the bird. Viewers can also determine if they have been sent in an obvious direction (clockwise or counterclockwise), a decision of the artist so that viewers will consider some things before others. Viewers are likely to enter the picture as though accompanying the honey hunter and then follow his gaze to the hive and to the retreating bees. They may return then to the honey hunter or move up to the perched honeyguide. Routing viewers through a narrative can be aided through various artistic devices— the use of light, space, negative space (emptiness), color, perspective, detail (lots or a little), paint as a texture, and so forth. Here, a dark foreground opens up to a lighter background, negative space and positive space are relatively balanced, and the number of elements is minimal.

PLATE 41. *Twittering Machine (Zwitscher-Maschine)*, 1922, 151, by Paul Klee

PLATE 42. *Drawing Based on a Detail from Paul Klee's "Twittering Machine" with Sonogram*, 1995/2007, by Darryl Wheye

PLATES 41 AND 42 **Telling a Story about Birds and Technology**

The Narrative

The role of the imagination is readily seen in Paul Klee's *Twittering Machine,* but it takes a caption to point out interesting overlaps of science and art. The evocative caption that accompanies New York's Museum of Modern Art online presentation of the *Twittering Machine* describes the relationship between birds and machines: "The 'twittering' in the title doubtless refers to the birds, while the 'machine' is suggested by the hand crank. The two elements are, literally, a fusing of the natural with the industrial world. . . . . Each bird stands with beak open, poised as if to announce the moment when the misty cool blue of night gives way to the pink glow of dawn. The scene evokes an abbreviated pastoral—but the birds are shackled to their perch, which is in turn connected to the hand crank.

"Upon closer inspection, however, an uneasy sensation of looming menace begins to manifest itself. Composed of a wiry, nervous line, these creatures bear a resemblance to birds only in their beaks and feathered silhouettes; they appear closer to deformations of nature. The hand crank conjures up the idea that this 'machine' is a music box, where the birds function as bait to lure victims to the pit over which the machine hovers. We can imagine the fiendish cacophony made by the shrieking birds, their legs drawn thin and taut as they strain against the machine to which they are fused."[11]

The bird on the left, the one with a tail, is not shackled and might represent a real bird, but we suggest an alternative interpretation of Klee's mechanistic view of song production.

Viewing the Science

This expressionist painting hung for five years in the National Gallery in Berlin before Adolf Hitler proclaimed it degenerate. We see it not as degenerate but as imaginative and as a step ahead of science in portraying the "score" of a birdsong: we can view the heads of the birds as the points on a sound spectrograph representing the phrasing and frequency of a vocal exchange, or as perhaps the heads of musical notes (Plate 42).

Art critics interpreting this painting have sometimes considered it a contemptuous satire of laboratory science. John Adkins Richardson wrote that Klee's "picture is of a machine for producing birdsong; one turns the crank handle and the birds tweet grotesquely, jerking up and down on a sine curve."[12] On the other hand, it could be interpreted to represent a very real set of biologically significant relationships—a vocal exchange between parents and chicks, with the vocalizations depicted as well.

Perhaps, then, the painting captures an auditory experience and presents it visually. Plate 42 includes a sonogram—that is, the notation used in sound spectrographs. It is possible, reading the birds from left to right, to determine the volume, intensity, degree of trilling, and degree of shrillness of their voices by the size, shape, and direction of the protruding tongues, seen by some as stylized exclamation points or arrows.[13] Klee's painting preceded the development of the sonogram—a scientifically representative "Twittering Machine"—by forty years.

PLATE 43. *Model of Flying Machine Driven by the Force of Man by Leonardo da Vinci*, c. 1505 / 1988, constructed by James Wink, Tetra Associates, London, based on Ms.B.f.74v and others

PLATE 43 **Scientists as Artists, Artists as Scientists**

The Narrative

Polymathic individuals, those who are competent in more than one field of study, may be at an advantage for making breakthrough discoveries—especially when they are able to draw upon both the arts and the sciences. "He who has once seen the intimate beauty of nature cannot tear himself away from it again. He must become either a poet or a naturalist and, if his eyes are good and his powers of observation sharp enough, he may well become both." Robert Root-Bernstein, who proposes that connection, drawing on the work of Konrad Lorenz, goes on to suggest that "polymathic ability must be a source of synscientific insight," where "synscience" refers to the resonance between nature and the total human being. Paleolithic cave-wall artists may have had just these qualities, and Leonardo da Vinci, whose remarkable accomplishments included creative works as an artist, architect, anatomist, engineer, stage designer, natural historian, geologist, and inventor, who painted thirteen undisputed masterpieces *and* devised enough contraptions to fill a museum, certainly did. Leonardo's studies of the mechanics of flight, including aspects of lift, turbulence, and air pressure, were based on direct observations of birds in flight. He wrote that the design principles of mechanical inventions would work if they remained consistent with the universal laws of dynamics: "The bird is an instrument operating through mathematical laws, which instrument is in the capacity of man to be able to make with all its motions."[14]

Viewing the Science

Leonardo's flying machine is certainly the work of a polymath, but the ambition to fly undoubtedly appealed to earlier polymaths as well. The quest is seen in the legend of Daedalus (see Plate 21) and his son Icarus. It is documented in the efforts of medieval European and Arab engineers and even included in the work of the thirteenth-century English philosopher Roger Bacon, who described a "flying machine in the middle of which a man could be seated and make an engine turn to activate artificial wings that would beat the air like those of a bird."[15] Leonardo scholars have found that drawings of the mechanics of bird flight and human-powered flapping machines to replicate it began to appear in his notebooks by 1488.

Leonardo's flying machines usually include a carriage with retracted landing gear to carry the flyer, which was based on what he saw in birds like the stiff-winged, quick-flying Swift (*Apus apus*). He wrote about the swift that it "is unable to lift itself directly from the ground since its legs are too short. When it starts to fly, it folds its *ladders* as in my drawing." One sketch (Madrid, I 64r) features a sphere-shaped carriage made from hoops of elmwood in which many passengers could stand. Another includes foot pedals that allow the flyer to operate two wings, but the most complete sketch for a flying machine (C.A. 341r) shows a hand- and foot-activated rotation system that allows both vertical and horizontal flight. Leonardo considered a number of systems to power his flying machines, including a spiral air-screw and a dragonfly-like arrangement with four enormous pilot-operated wings, but most of his drawings contain unresolved elements. There is no evidence that he personally ever got airborne using human-powered or mechanically powered devices, but his orchestrated failures eventually led him to conclude that human-powered flight ought to emphasize gliding, which takes advantage of air currents, with both wings and a rudderlike tail. In the end, his notes and drawings are more a series of questions than solutions.[16]

Without lightweight, durable materials, his efforts could not succeed. Leonardo's drawings served as the starting point for the model shown here, however, which was constructed 500 years later by James Wick. Wick's version uses the sort of materials that Leonardo had planned on. It weighs 650 pounds (300 kilograms). In comparison, *Daedalus 88*, the human-powered aircraft that crossed the Aegean in 1988, weighed approximately 69 pounds (approximately 31 kilograms).[17]

PLATE 44. *An Experiment on a Bird in the Air Pump*, c. 1768, by Joseph Wright of Derby

PLATE 44 **Social Links between Scientists and Artists**

The Narrative

Britain's Lunar Society did not include Joseph Wright of Derby (1734–1797)—the artist who produced this painting—although he could be thought of as an honorary affiliate, for its members, all of whom were interested in scientific discoveries, were friends or acquaintances. The society had been founded by Erasmus Darwin, Charles Darwin's grandfather, and was so called because its members met on the Monday nearest the full moon to allow members to ride home by moonlight. (Note the full moon through the window on the right.) Among its twelve principal members were Joseph Priestly (discoverer of oxygen); Matthew Boulton (industrialist and toymaker); Boulton's partner, James Watt (inventor of the steam engine); and Charles Darwin's other grandfather, the potter Josiah Wedgwood (see Plate 8). The group was social, not academic. Few had a university degree, although ten became Fellows of the Royal Society. Nor was it steadfastly Continental. It included, for example, William Small (Thomas Jefferson's mathematics teacher at William and Mary) and received occasional visits from Benjamin Franklin. Both the inclusion of Josiah Wedgwood as a member in the Lunar Society and Joseph Wright's production of this painting suggest that the science-art barrier was probably not significant at this point. The term "scientist" would not even be coined until 1830.[18]

Wright's painting features a traveling experimenter who visited the homes of men and women who had declared an interest in science and nature and hosted social events based on that shared interest. Here the experimenter is testing the effectiveness of an air pump against the respiratory needs of a Sulphur-crested Cockatoo (*Calyptorhynchus galerita*) sealed in a glass chamber. The cockatoo is sinking. Is it going to die as more air is pumped out of the glass flask? Its fate is up to the visiting scientist, whose hand grips the stopcock at the top of the flask. The artist was not trying to dazzle his audience by showcasing an air pump as a new device, for it had been around for a century. Otto von Guericke, a German physicist, engineer, and natural philosopher, had invented it in 1650. In fact, familiarity with the pump was essential if Wright's dramatization of the life-or-death scene was to work.[19]

The evident interest of a lay social elite in a staged biological experiment is perhaps even more remarkable than the mixed scientist-artist membership of the Lunar Society. Also around 1768, Wright painted another narrative of drawing-room science. This time the subject was astronomy. Together, the air-pump painting and *A Philosopher Giving a Lecture on the Orrey* show two central aspects of science (experimentation and teaching) in a dramatic and appealing way and convey the excitement of the discipline. Engravings of both were made, and the resulting prints were very popular.[20]

Viewing the Science

Normally a sparrow, lark, or other common songbird was used as a "guinea pig" in demonstrations and experiments (see Plate 22). Either the valuable, exotic cockatoo was readily available and there was no intention of harming the bird, or Wright was exercising artistic license. Wright's use of a passerine in the preliminary study for the painting points to the latter. In either case, if those watching were holding their breath, they would have concluded that the bird was in greater danger than it actually was; they would have run out of air before the bird did, because birds have a slower respiratory rate than mammals do. Their lungs are not pliable, like ours, and do not work like a bellows, taking their breath in and pushing it out. Their breathing cycle is more complicated and more efficient: they are fitted with a system of air sacs that allows them to breathe in and out at the same time.

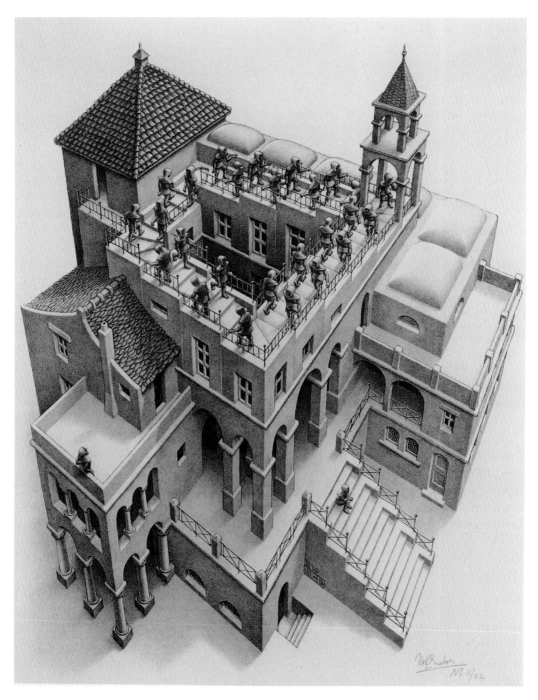

PLATE 45. *Ascending and Descending*, 1960, by M. C. Escher

PLATE 45 Collaborations between Scientists and Artists

The Narrative

When a scientist and an artist collaborate, the art can change the viewer's standpoint. The physicist and science writer Alison Boyle has written about Roger Penrose (b. 1931), known for his work in relativity theory, quantum mechanics, and mathematical games, who invented, with the aid of his father, the Penrose Staircase, which unravels the laws of perspective. We can see that staircase in M. C. Escher's lithograph *Ascending and Descending*. How it got there is described in a letter from Escher to Penrose dated April 18, 1960: "A Dutch friend of mine sent me some months ago a phototype of your article.... Your figures 3 and 4: 'continuous flights of steps' were completely new to me and I was so impressed by the idea, that it inspired me recently to [make] a new print, of which I should like to send you, as a homage, an original copy." He went on to ask for articles by Penrose or anyone else "about 'impossible objects' or allied subjects with good figures."[21] Penrose apparently sent more material.

Viewing the Science

Three years after Robert Root-Bernstein's 1995 study on the "Two Cultures" divide (see the discussion in the Mezzanine), Caroline Jones and Peter Galison analyzed how images function in art and science. In their view, science is more than "revealed Truth," and art is more than "individual statement[s]." Both science and art create culture, and neither is independent of it, so context matters: science cannot rise above society any more than artistic genius can direct it from somewhere beyond its influence. Jones and Galison also support the view that the boundary between science and art is weak and that when evaluating images, historians—of science, of art, of philosophy, and of culture—will benefit from considering the conditions at the time they were produced.[22] We believe that viewers would benefit as well. Recognizing the Penrose-Escher connection, for example, allows a richer assessment of both the Escher image and the Penrose Staircase than would be likely otherwise.

To Marshall McLuhan's assertion that "art at its most significant is a distant early warning system that can always be relied on to tell the old culture what is beginning to happen," we would add "and the consequences of the choices that lie before us."[23] The art-science disciplinary convergence can serve to alter the viewer's perception of both the objects portrayed in an image and the science underlying their relationship. The authors of scientific articles about nature strive to eliminate ambiguity so that when each reader gets the point, they get the same point. That tradition of exactitude disciplines the scientists, conserves credibility, and helps broaden the base of knowledge that supports prevailing theories. Artists painting slices of nature—when the subject is well documented in scientific articles—may, in contrast, insert some ambiguity so that more than one message or conclusion is available for viewers to extract. This multiplicity of viewpoints keeps art alive and makes it possible for a viewer to revisit the work and hope to find a new interpretation or a new message. But if one among a host of possible interpretations stands out, most viewers are unlikely to miss it. Thus, art illuminates science, and science inspires art, and each serves the other to our benefit (see, e.g., Plates 30, 44, 50, 53, and, of course, this plate by Escher).

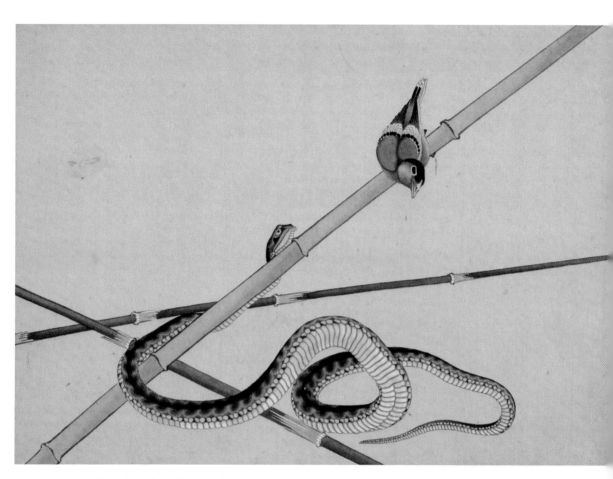

PLATE 46. *Snake and Small Bird*, c. 1836, by Hokusai

PLATE 46 Visual Metaphors That Are Grounded in Science

The Narrative

Hokusai reduces the elements in this painting to the snake, the bird, and precariously poised bamboo, inviting us to become involved in the tense balance between a predator and its intended prey.[24] The snake contorts its body as it approaches the bird. If it moves too far or too fast, the bamboo will become unstable and bird will take flight, but if it does not move in time, the bird will fly away anyway.

Viewing the Science

The notion of balance can be seen through the lenses of both science and art. Some viewers might be drawn into this narrative because of the visual equilibrium. Some might be drawn in by the metaphor of maintaining an ecological balance between predators and prey. Others might interpret the painting as a metaphor for the evolution of birds from their reptilian past, even though birds are descendants of lizards rather than snakes.

Naturally, the fusion of science and art has the capacity to dissolve the boundary between the two. But there are other boundaries. One boundary lies between us and the work itself: we can be put off by the style or the medium. Another lies between our personal experience and the story we find unfolding. A third lies between rationality and emotionality as these coexist and contest with one another in our thoughts. Their dissolution also reduces our resistance to entering a work and identifying with the relationships it displays.

Viewing significant relationships in nature helps us understand the kinds of interactions between organisms that generate ecological structure. If the work convincingly represents the essence of one interaction, we may suspend our disbelief and ignore that we are looking at a canvas covered with paint or, in this case, a piece of paper coated with pigment—that is, we momentarily experience the scene as though it is real and we are there. A great pictorial narrative can get us involved, even if we cannot identify fully with, say, a snake. When it does, our sense of nature can sharpen, and our understanding of our place within it can deepen. If the narrative is persuasive, we might even shed some long-held but unexamined assumptions.

The aesthetic component that draws artists into portraying nature may resemble the aesthetic component that draws scientists into observing nature. When scientists write or talk about breakthroughs, they sometimes invite the rest of us to imagine the fervor of the moment of discovery, a human aspect of science that is usually held at bay in the effort to keep experimentation and analysis "objective."[25] Hokusai, who produced more than 30,000 drawings and paintings over his seventy-year career—but died in poverty—could be inviting us to imagine more than the plight of the snake and the bird.

PLATE 47. *Muscles That Raise and Lower the Lower Mandible,* 1984, by Tony Angell

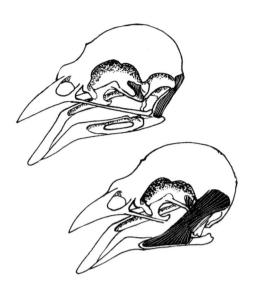

PLATE 48. *Male Black-backed Oriole with Captured Wintering Monarch Butterfly in Pine Woods above Mexico City,* 1984, by Tony Angell

PLATES 47 AND 48 **What Quantity of Information Says about a Work of Art**

The Narrative
All art illustrates, but not all illustrations, even if they are imaginatively rendered, are art. The diagram of two bird skulls in Plate 47 is clear, and viewers might mentally map them onto birds whose foraging habits they know. Plate 48, however, a pen-and-ink drawing of a male Black-backed Oriole (*Icterus abeille*) grasping a wintering Monarch butterfly (*Danaus plexippus*), not only conveys biologically instructive information about the bird as it feeds, but also engages us. As the artist, Tony Angell, notes: "When you can utilize the aesthetic to catch the reader's eye and still share the information, you're ahead. You've reproduced a truth from nature through your creative imagination. The content of the composition is at once accurate and engaging. I wanted to connect the two attractive patterns of the bird and butterfly while they are surrounded by the delicate filigree of needles in the tree they both occupy."[26]

The artist proved his familiarity with the bird's behavior through his selection of the picture's content, and he exercised his craft through his selection of medium. Ink is unforgiving, but Angell finds that using it enhances the drama of a picture, making it more compelling. When successfully used, this challenging medium adds a level of authenticity to a depiction. Moreover, it reproduces well (and relatively inexpensively) on the printed page, letting its aesthetic appeal and technical finesse be readily appreciated.

Viewing the Science
The first of these two examples from *Blackbirds of the Americas*, illustrated by Tony Angell, accompanies a description of gaping (forcing the bill open against resistance).[27] The line drawing helps viewers visualize the mechanism that allows a bird to pry apart a small crevice so that it can retrieve hidden food (Plate 47). The upper part of the drawing identifies the muscle in the back of the skull responsible for depressing a bird's lower mandible (allowing the bill to gape), and the lower part of the drawing identifies the much larger muscle on the side responsible for raising the lower mandible (allowing the bill to crush). Together, the upper and lower parts clarify and amplify the associated text and for most people are secondary to it. Even readers who are quick to comprehend images gain confidence in their interpretation by reviewing the text itself, for there they learn that the muscles for opening the bill are bigger in birds that do a lot of gaping and smaller in birds that do a lot of crushing, which would make the bird in the drawings basically a crusher.

In contrast, the drawing featuring an oriole with a butterfly in its beak enables the viewer to process far more information without the need to read a lot of explanatory text (Plate 48). Someone looking at it can identify the predator, its prey, and their habitat, make some assumptions about the bird's previous activity, and anticipate what might happen next. It is evident from the pristine condition of the butterfly that its capture was neatly and recently done. It may be assumed that the pine bough, dipping from the weight of the bird, will sway slightly when the bird departs for a perch where it can de-wing the butterfly and eat it at leisure. Almost any reader would quickly classify the figure of the skulls as an illustration, but many would see the drawing of the bird as art—and many more would if it were placed in a $500 frame and hung on the wall of a penthouse apartment.

PLATE 49. *Woman Observing Bird (Willie Was Different)*, 1967, by Norman Rockwell

PLATE 49   **What History Says about Classifying a Work of Art**

The Narrative

Streamlined illustrations showing the essence of a subject have existed since cave art and have played a large role in the history of scientific images. Under many circumstances, these images are viewed as art, however, not as diagrams or "mere" illustrations, and people who produce them are viewed as artists. Is the owl from Chauvet Cave art or illustration (see Plate 2)? Was Norman Rockwell an artist or an illustrator? How about Roger Tory Peterson? Conventional wisdom may say that both were illustrators. But conventional wisdom can oversimplify. For four decades Norman Rockwell's pictures were published on the cover of the *Saturday Evening Post,* and other Rockwell images appeared in advertisements or accompanied stories. So he could be called an illustrator—but he had the largest audience of any living artist in history.[28] And for even longer, Roger Tory Peterson's pictures illustrated his field guides, but his paintings hang in many museums.

Rockwell's *Woman Observing Bird* was the cover illustration for a children's book, *Willie Was Different,* cowritten with his wife, Molly Rockwell. The satirical fantasy is about a rebellious thrush that rejects the traditional thrush song for the notes of a flute he hears through an open window. The story is a thinly veiled commentary on the creative spirit that exhausts itself seeking happiness through lasting fame, for the bird dies of exhaustion, but his song is recorded and a room is devoted to him at the Smithsonian Institution.[29]

Viewing the Science

Like Paleolithic cave-wall art, this image lacks a background. Nevertheless, in the minds of many it is an illustration and not art. On the other hand, even though the image blends fiction and reality, the source material is authentic, and the elements —a musical bird, a musical instrument, and an observer—are logical. Rockwell based his thrush paintings on a specimen prepared by Ralph C. Morrill of Yale University's Peabody Museum. He based the figures in a related image, which includes the dignitaries who recorded the bird's song, on Morrill; S. Dillon Ripley, the secretary of the Smithsonian Institution; and Philip Humphrey, a trustee of the National Audubon Society, each of whom sat for the artist.

The Rockwells' choice of a thrush is biologically sound. In consultations with ornithologists and musicians they had learned that works by Frederic Handel, Carl Maria von Weber, and Charles-François Gounod all contain passages inspired by the thrush. The authors' point about song appropriation is biologically sound, too. Some birds are known to incorporate musical scores into their repertoires. Wolfgang Amadeus Mozart's pet European Starling (*Sturnus vulgaris*), like others of his species, was a great mimic who appropriated the composer's themes, including the beginning of the last movement of his Piano Concerto in G major, K.453, which was written about the same time the bird was purchased (May 27, 1784).[30]

Through a blend of fiction and reality, and of image and text, the Rockwells brought birdsong and bird watching into the lives of a generation of North American children. The illustration was published a decade before President Gerald R. Ford awarded Rockwell the Presidential Medal of Freedom, the highest civilian honor. He was honored as both an illustrator and an artist.

PLATE 50. *Brewer's Blackbird*, c. 1914, by Louis Agassiz Fuertes

PLATE 50 **What Venue Says about a Work of Art**

The Narrative

Louis Agassiz Fuertes' watercolor painting of a Brewer's Blackbird (*Euphagus cyanocephalus*) eating an Alfalfa Weevil (*Hypera postica*) was produced for the *Bulletin of the U.S. Department of Agriculture* in 1914 and was published as one of five plates in a report by E. R. Kalmbach on the results of his study in north-central Utah in 1911 and 1912 that sought to correct the misconception that the bird eats crop seeds rather than insect pests. The Alfalfa Weevil had been introduced into that part of the country in 1904, and Kalmbach's study showed how forty-five bird species adapted to the new food. The blackbirds and the sparrows were the most effective in controlling the pests. Fuertes signed the watercolor, and the *Bulletin* listed his name beneath the illustration, but nonbook reproductions of the image may not include a credit and may publish the image at a size that makes reading the signature very difficult.[31]

From the 1920s to the 1940s, Fuertes brought images of North American birds into countless households in the form of small cards inserted into boxes of Arm and Hammer baking soda. He signed the watercolors, but his name was not typed on the cards. Nonetheless, this venue for his illustrations helped to make Fuertes the best-known North American bird artist, recognition that he acquired despite having written very few articles and no books.[32]

Was he an artist or an illustrator? Whether the artist of a work is identified with the work gives one clue to perceived status. Whether the work is supplementary to a text is another clue. How credit is given is yet another—whether just in the acknowledgments, say, or more fully with the art, along with the title of the art, the medium, the size of the image, and the location of the art, plus perhaps other details. But the clues cannot be definitive, and they are often contradictory. In fact, the distinction is arbitrary. All the works in this book are both illustrations of Science Art and works by artists.

Viewing the Science

Fuertes' blackbird was later included, along with thirty-five other watercolors and thirty-six black-and-white drawings by him, in the posthumously published 1,100-page *Bird Life of Texas,* by Harry Church Oberholser. Oberholser worked for the U.S. Bureau of Biological Survey (later the U.S. Fish and Wildlife Service) from 1895 to 1941. When the bureau sent him to Big Bend, Texas, to survey birds and mammals, Fuertes went with him. Fourteen years later, Fuertes painted the blackbird. According to Fuertes' daughter, Mary Fuertes Boynton, he was happy for the work even though he believed that "pictures of birds are just about the last things anyone needs in war time." But this is not just a picture of a bird. It tells an important story about the value of the bird's help in ridding the field of weevil larvae.[33]

Interestingly, *The Bird Life of Texas* lists Fuertes' name above that of E. B. Kincaid, the editor, and the listing on the title page reads "Paintings by Louis Agassiz Fuertes," not "Illustrations by Louis Agassiz Fuertes." Here is an example of the way a shift in venue (painting to book) that ordinarily would produce a shift in category does not. It adds credence to the assertion that the division between art and illustration is arbitrary.

The artist Robert Bateman (see Plate 36) sums up the disservice to art labeled "illustrations" this way. "A book publisher asks N. C. Wyeth to illustrate Robert Louis Stevenson's 'Kidnapped' or Pope Julius II asks Michelangelo to depict the story of creation on the Sistine Chapel ceiling. Pure, non-objective painting, I presume, would be 'real' art because it is about itself. And so a furniture store abstract that matches your drapes is real art and da Vinci's 'Last Supper' is a *mere* illustration. It is the word 'mere' that I find troubling."[34]

ROOM 7

Content, Style, and Medium

The ten bird images in this room show some of the influences that content (narrative) and the artist's style and choice of medium have on Science Art. The first five exemplify the role of content; the next two show two artistic methods; and the last three present three media—photography, realist painting, and minimalist serigraphs.

Content

When nature artists invite us to take a closer look into an animal's life, what is presented is as accurate as the artist can make it. This does not mean that each blade of grass will be in place, but if the artist puts the subject within a grassland habitat, the scene will reflect the qualities that give that ecosystem its characteristic look and feel. Artistic license cannot extend to the point of misrepresenting nature for the sake of aesthetics, but it *can* give the viewer an otherwise impossibly close look at nature and the influence of human activity on animals and plants and their habitats.

The first three paintings in this room feature Common Ravens in very different narratives that portray aspects of their solitary, intraspecies, and interspecies behavior. The next two feature Evening Grosbeaks in narratives that convey aspects of their ecology and the influence of human activity on their population. Humans have altered the ranges of these two very different birds. Both have been increasing in North America; the former through our commensal relationship, the latter with the help of our backyard feeders (see Plates 52 and 54).[35]

Style

As we saw in Room 6, distinguishing between illustration and art is an arbitrary and essentially meaningless exercise. Whether we classify an image as an illustration or as art is partly a matter of perception (opinions vary among viewers), partly a matter of convention (cultural attitudes change with time), and partly a matter of venue (impressions change with context), so published art, for example, is often referred to as illustration. The choice also depends in part on who controls its production. The artist paints a picture of his or her choosing. The commissioned illustrator agrees to create what is required to accompany the text in a publication, an advertising campaign, or a label. Artists may endure the assignments to eventually achieve relative autonomy; during this phase in their career they may prefer not to think of themselves as illustrators, even if that role serves as the basis for their later success. For seven decades Roger Tory Peterson's field guides taught tens of millions of us how to identify birds. Since far more people have seen his illustrations in his field guides than have seen his art in museums, it is not surprising that he is known to many as an illustrator. But we can also see him as he saw himself: as an artist.

Nor can we always distinguish between photo-realism and art produced using a more painterly style. As the name suggests, an artist using photo-realism renders an image in such meticulous detail that it has the attributes of a photograph. An artist using a painterly style generally creates a more open composition, often forming shapes through variations of color rather than through outlines or contours, so the painting of a bird looks like a painting, not a photograph. But even when comparing styles as different as these, complications surface right away. Within a single image there may be elements of both, forming a hybrid that is likely to appear primarily as one or the other (see Plate 57).

In two images of the same subject there may be differences in the amount of information conveyed, especially if one image excludes details. Because artists are selective—they have to be: it is impossible not to make choices as they make art—they may choose to incorporate only parts of an original scene or to eliminate a degree of detail—as, for example, when they choose a large brush to make rough, sweeping strokes instead of a small one that could capture minutiae. Or they may choose to elaborate by adding elements to a scene. Such variation in technique, details, and degree of reality, along with differences in the perception and knowledge of both the artist and the viewer, help explain why style can be difficult to pin down. But these variations also explain why the influence of style on content can be so significant—and why it requires, from the viewer, a certain tolerance for imprecision.

Sometimes a piece of art can leave us at a loss to describe its style. Some artists have a tendency to develop highly personal styles, making labeling difficult. But we can keep the matter of style simpler and more understandable if we discard any reference to it in historical terms, except where that may be useful, and focus instead on some fundamentals of technique.

Researchers are chipping away at some of the difficulties in understanding the influence of style on the way we look at art; they seek clues by measuring the visual perceptions of both the viewer and the artist. In one rather well-known study, the neurobiologist Margaret Livingston analyzed the effect of the viewer's focal point on the size of the Mona Lisa's smile. She found that when viewers used their peripheral vision, that is, when their gaze shifted from her mouth to her eyes or hands, she seemed to be smiling more. In another study of Leonardo's iconographic painting, researchers analyzed three-dimensional images produced from extremely high-resolution laser and infrared scans. They were able to see under the surface paint that her hair formerly had a different style and her hands were in a different position (they gripped the chair). They were also able to detect a gauzy overdress, traditionally worn by nursing mothers in the sixteenth century, which had been obscured by darkened varnish. With this new evidence, the identity of Mona Lisa was finally discovered: she was Lisa Gherardini, a new mother. Now that we know who the Mona Lisa appears to be and can guess at her reason for smiling, it would be interesting to find out whether knowing alters viewers' perception of the smile. In a third study Livingston, along with Bevil Conway, found that a vision abnormality might have helped shape Rembrandt van Rijn's art. They analyzed his thirty-six self-portraits and found that all but one indicates that he had divergent strabismus, or stereoblindness, a condition that would have inhibited his ability to see in three dimensions. If he really was walleyed, objects would have seemed rather flat, and shadow and light would have taken on considerable significance.[36]

Artists using a photo-realistic style provide clarity, but they are not themselves cameras. A successful depiction will depend in part on their having selected the elements that approximate the feel of the biological situation they are trying to capture.[37] This mix of realism and arbitrariness may trouble some, but images may contain at least as much information as a written account by a scientific researcher, and the bias of the artist does not necessarily exceed that of a writer describing the same thing. Just as there are better and worse scientific accounts, so are there better and worse examples of Science Art, including the work of artists who successfully use a painterly style. These artists capture our attention by employing an aesthetically forceful technique that persuades us to look past brush strokes and vagueness and to make our assessments based on only a fraction of the elements that would have been present in nature. As a result, we may feel rewarded by seeing essential elements revealed, or we may perhaps regret not having more details—

the way we wish for a second helping when the chef is good but the serving meager.

Styles are juxtaposed in two paintings in this room: in one we see the realism of Ray Harris Ching and in the other, the painterly approach of Thomas Quinn. The subject of both paintings is the act of waiting, a requirement that we share with birds even though we sometimes lack their patience. In Ching's assembly, all but one monkey and the dove are waiting by sleeping (see Plate 56). In Quinn's double portrait, one bird waits calmly; the other takes a momentary break (see Plate 57).

Medium

There is merit to the advice once given to those interested in memorializing nature: If your subject is nature and you want an image to accompany your text, use a photograph. But the photographic approach comes with a number of limitations.

First, some photographs are impossible to take. The last Passenger Pigeon, Martha, died in the Cincinnati Zoo in 1913, so the extraordinary flocks of passing Passenger Pigeons, the magnitude of which impressed Audubon a century before Martha exhaled her last breath, were gone before cameras were commonly available. All we have to refer to are descriptions like Audubon's: "Feeling an inclination to count the flocks that might pass . . . in one hour, I dismounted, seated myself on an eminence and began to mark with my pencil, making a dot for every flock. The birds poured in in countless multitudes, I rose, and counting the dots then put down, found that 163 had been made in 21 minutes."[38]

The camera dates to 1822, so the world's archive of nature photography is less than 200 years old. Not surprisingly photography was very sparse early in its history. An image showing what Audubon saw in Kentucky that day in 1813 requires an artist.

Second, some great photographs were never taken because the opportunity was lost. We must also respect the difficulty of taking a truly great photograph of birds in motion; the number of those in the world's archive is very much smaller. For *Winged Migration* (2003), the superb, Oscar-nominated film, 95 percent of the footage (from which still images were made) was left on the cutting room floor. That is, the usability figure was 5 percent. This compares with 2–4 percent for a typical wildlife film and 10–20 percent for feature films.[39] So although the photographic archive is massive, the percentage of memorable photographs is very small indeed.

In recent years, however, innovations in cameras and in the vehicles carrying them have increased the odds of getting great shots. Digiscoped images (taken through a spotting scope with a digital camera) and images taken with digital single-lens-reflex cameras are making a significant difference in overall image quality. At the same time, cameras mounted on various ingenious carriers are improving the chances that we can record birds in motion. To capture the essence of migratory flight, the filmers of *Winged Migration* used helicopters, paragliders, gliders, paramotors, delta wings, balloons, remote-controlled mini gliders, mini helicopters with remote-controlled cameras, ultralights, zodiacs, boats, ships, and amphibious robots.[40]

The last three images in this room are of curlews. Comparing them allows us to explore some of the advantages and limitations of photography, of paintings made with the aid of photographs or a photographic memory, and of images that fit under the rubric of minimal realism.

PLATE 51. *The Raven*, 1995, by Henry Bismuth

PLATE 51 **Content: Revealing an Individual's Behavior**

The Narrative

This portrait of a Common Raven (*Corvus corax*) demands our attention. The bird calling out to an unseen companion seems as insistent as Stanley (played by Brando) calling out to an unseen Stella. The portrait is accurate: those familiar with ravens recognize the distinctive posture, feathering, and coloration. It is convincing: those familiar with ravens all but hear the raspy croak. It encourages a closer look—for viewers sense that the bird is straining to communicate—and it involves viewers by awakening curiosity: What is the raven communicating and to whom? The aesthetic elements are striking and draw viewers in.

The artist, Henry Bismuth (b. 1961), is a physician residing in Gentilly, France. He wrote the following caption for this painting: "My interest in corvids began a few years ago with a sudden overpopulation of these species in my region. Their presence allows me a wide range of pictorial explorations. I like the shape and colors of ravens, but most of all I like their behavior and spiritual presence in various cultures around the world. This is the largest painting I have ever done. The raven is depicted on a human scale, and I very freely mixed abstraction and realism to better express an emotional approach to the subject."[41]

Viewing the Science

In *The Mind of the Raven* the biologist Bernd Heinrich describes both semidomesticated ravens in Maine and a wild flock he has studied intensively enough to identify individuals. He makes a persuasive case for the capacity of these birds to recognize one another and act appropriately, that is, in a history-dependent fashion. He has also developed an interesting and convincing case that there has been a historic coevolutionary behavioral relationship between ravens and wolves, their fellow residents in the boreal forest. Indeed, corvids have big brains—the largest, relative to body size, of any bird. They also have big vocabularies. Dozens of vocalizations have been described in the ornithological literature, and the birds are thought to vary meaning by modifying intensity, pitch, sequence, and repetition rate.[42] Thus, our sense that Bismuth's raven is doing more than advertising its presence—the way a songbird would during a dawn chorus—seems warranted.

Some of these big-brained corvids—crows, ravens, and magpies in particular—are currently the primary avian victims of the West Nile virus. It seems likely that they will increasingly be the subject of paintings featuring birds, for images help us adjust to the presence of this new, dangerous virus, its avian host, and its mosquito vector. The Cornell ornithologist Kevin McGowan described a flock of crows that he had been monitoring when it was struck by the West Nile virus: "Healthy crows stood sentinel over dying companions, and one widowed bird took her children and moved back in with her parents."[43]

PLATE 52. *Picnic*, 1997, by Paula G. Waterman

PLATE 52 **Content: Revealing Intraspecies Behavior**

The Narrative
One of the Common Ravens (*Corvus corax*) in *Picnic* is setting out to devour a food bonanza; the other is possibly momentarily attentive to the need to serve as a lookout. The painting is accurate and convincing, from the heavy, almost iridescent plumage and erect neck feathers of the rear bird to the stare-down gaze of the front one. The gluttonous intensity and watchfulness invite a closer look. Viewers who pause to inspect the painting might look for clues to determine whether a person left the trash behind, whether an animal retrieved it from a trash bin—or, if the birds had managed to retrieve it, how they dealt with such a large object. The aesthetic elements of the painting—the diagonal construction, the glossy feathers, the crinkled foil and plastic wrap, all brilliant against the flat dullness of the dirt and cracked rock—also draw viewers in.

The source of light is a key element in the work of the Maryland artist Paula Waterman. Light is clearly seen here: it ricochets off reflecting surfaces and picks out the intricacies of feathers. Waterman wrote the following caption for this painting: "The raven is big, surprisingly colorful, and tends to dominate any scene. *Picnic* can refer to the birds' actions or the actions of those who left the food behind. It is not meant to be a political statement; garbage is simply a fact of life. Some birds take advantage of its existence."[44]

Viewing the Science
Ravens are omnivorous. They seek fresh meat but will scavenge for carrion, prey on eggs and nestlings of other birds and the occasional arthropod, take seeds and grain, and pick through garbage. Since raven populations are increasing in areas where humans leave garbage, a commensal relationship between these birds and people appears to be developing. As noted earlier, in commensal associations, members of one species assist the members of another but incur no significant costs and receive no benefits from the relationship. Here, however, humans may be benefiting from the ravens' free service as a cleanup crew.[45] Ravens may also have a commensal relationship with wolves (see Plate 51).

PLATE 53. *Without Warning*, 1998, by Carl Brenders

PLATE 53 **Content: Revealing Interspecies Behavior**

The Narrative

This painting of a Common Raven (*Corvus corax*) and a Golden Eagle (*Aquila chrysaetos*) contains all of the elements of a Hollywood chase scene without the metal and the mess. Dexterity and cunning are pitted against size and power. The presentation is accurate: the raven, perhaps one of a group, is mobbing the eagle—trying to get it to fly away—and we can all but hear its scolding call. It is also convincing: viewers can sense the weight of the eagle and guess how few wing beats could move it beyond the raven's reach. The painting encourages a closer look because of the sense that the eagle is ready to act. Those who are familiar with such chases will expect the raven to prevail and the eagle to leave. The question is whether it will leave peacefully. The aesthetic elements— the forms, the feathers, the balance of positive and negative spaces, the balance of light and dark, the balance of tension and relaxation, draw the observer in.

The Belgian artist Carl Brenders wrote the following caption for this painting: "*Without Warning* creates the feeling of being high in the sky with the most beautiful flying creature in the world—a Golden Eagle—surrounded by an evening sun penetrating the mist. . . . I cropped the birds' wings to strengthen the feeling of being close to the action and part of a tension-filled moment in the sky."[46]

Viewing the Science

Mobbing is an antipredator, anti-intruder strategy in which one or more birds chase or harass a larger one. This David-and-Goliath behavior arises when a patrolling avian predator alerts the neighborhood watch, and resident birds (alone or in a mob) try to chase it away. They will approach hawks, eagles, owls and sometimes nest-robbing species like ravens, crows, and gulls that come into their area. Sometimes small birds gather to mob an owl caught resting during the day, and some birders even carry tape recorders with owl calls to encourage mobbing, which may make the resident birds easier to see.

Mobbing tends to occur most intensely on the mobbers' breeding territory, although in some cases the mobs are large and include even distant neighbors as well as the territory holders. Thus, one of its functions is to divert the predator or intruder from areas in which there are fledglings by confusing or annoying the target bird enough for it to move away. Birds may learn from one another which birds to mob, and a gathering crowd of mobbers alerts others to the presence of a potential threat. The original mobber benefits directly if the predator leaves—and even if the mobber is attacked and killed, its behavior may protect its kin. In any case, mobbing is not as dangerous as it looks, although the behavior is not risk-free. Predators who hunt in pairs may provoke mobbing as a hunting strategy (the second bird swoops in for a kill), and predators who eavesdrop on mobbing birds may zero in on active nests.

PLATE 54. *Evening Grosbeak*, c. 1948, by Roger Tory Peterson

PLATE 54 **Content: Revealing Community Ecology and Habitat**

The Narrative
This rendering of Evening Grosbeaks (*Coccothraustes vespertinus*) was painted by Roger Tory Peterson, an artist, as noted earlier, best known for his field guide illustrations. Interestingly, some of his gallery art incorporates aspects of his field guide illustrations that portray nature. These Evening Grosbeaks, for example, accurately display dorsal, ventral, perched, and airborne positions, allowing viewers to map what they have seen in the field or in the field guide and readily recognize the species from all angles. Yet it also convincingly conveys the swiftness of flight—despite the frozen poses—and the chill of snow mantling the berrylike fruit. It encourages a closer look, for viewers sense that the birds must depend on these berries to augment whatever else they can find to eat in a snow-encrusted landscape. Some may wonder whether the berries are a widely available resource or merely a hedge against starvation. Others might ask if the berries have begun to ferment. The aesthetic elements—the forms, the diagonal construction, the balance of positive and negative, light and dark, and motion and chill—all these draw viewers in while they ponder the content.

Viewing the Science
Little is known and less has been written about Roger Tory Peterson's lithograph of Evening Grosbeaks, but it is of interest for reasons having nothing to do with its provenance.[47] When Peterson published *A Field Guide to the Birds* in 1934, he introduced identification techniques that virtually anyone could master, and in untold numbers people did. Many also did something else: they put food out for birds in the winter. It has been said that Peterson is partially responsible for expanding the range of the Evening Grosbeak (and the Northern Cardinal, Tufted Titmouse, and House Finch, among others). Prior to 1890, Evening Grosbeaks, who are short-distance migrants, were unknown east of the Great Lakes. Over the next two decades they became casual wanderers to the eastern United States, and between 1910 and 1932 they became more or less regular winter visitors in small flocks of six or eight to as many as a hundred birds. Individuals banded in Michigan were taken in Massachusetts and Connecticut, showing that they came from the Midwest. By 1959 the birds were breeding in the spruce belt of northern Michigan and spending the winter (erratically) in Missouri, Kentucky, Ohio, Virginia (rarely), and New England. By 1977, as the National Audubon Society reported in their field guide, "This grosbeak formerly bred no farther east than Minnesota, but more food available at bird feeders may have enabled more birds to survive the winter, and the species now breeds east to the Atlantic."[48]

The birds have continued to expand their breeding range in the eastern United States.[49] But they are being closely watched: the impact of climate on community ecology is a topic of increasing interest as the effects of global climate change on bird populations are increasingly evident. While researchers are exploring bird biogeography, "citizen scientists" (a term coined by Rick Bonney at the Cornell Laboratory of Ornithology in the 1990s) are participating in feeder watches and other studies and collecting data to help conservationists assess the changes in avian populations. In addition to the annual Audubon Christmas Bird Count, citizen scientists also participate in the annual Great Backyard Bird Count, sponsored by the Cornell Laboratory and the Audubon Society, which is expanding to include parks, refuges, and schools. For the record, Roger Tory Peterson, who served as the Audubon Society's education director and the art director for its magazine, compiled 158 counts and participated in 345. Also for the record, reports show recent declines in numbers of Evening Grosbeaks, attributed to global climate change.[50]

PLATE 55. *Champion,* 1998, by Patricia Pépin

PLATE 55 **Content: Revealing the Outcome of Human-Animal Interactions**

The Narrative

This painting of an Evening Grosbeak (*Coccothraustes vespertinus*) shows what can happen when the techniques of photo-realism pull us into a scene that we fully believe is a real place. Here we meet a bird at eye level and view the landscape as it is reflected back from the fender and hubcap.

Those familiar with Evening Grosbeaks can immediately identify the bird, and those familiar with Studebakers will surely recognize the Champion. The painting is so convincing that the bird seems ready to hop from its perch at the first sign of danger. The realistic rendering encourages a closer look as viewers try to decipher the reflections in the metal for more information about the locality, the car, and the bird. (The tires are modern narrow-banded whitewalls, not the broadbanded ones that would have come with the car originally, and they are clean, so it looks as though the car may still run.) The aesthetic elements—the truncated car, the palette, the bits of yellow that move the eye around the composition, the balance of bright and dull, light and dark, massive and fragile, immobile and flighty, and inanimate and living—all invite viewers to consider the content. The painting suggests multiple themes of obsolesce and senesce, self and other, territory holding and perceived threats to territory, and avian habitat and the human footprint on it.

This portrayal is testimony to the passage of time. The Canadian artist Patricia Pépin has a great appreciation for natural cycles and life spans: "Nature breaks down things all the time, through weather and microorganisms. Everything is being digested all the time, and nowhere is this more evident than on manmade things." The invasive nature of decay is more apparent upon scrutiny of the car she found "behind a gas station in Maine, with many other vintage cars, in different stages of decrepitude, destined to be restored or salvaged for pieces." Pépin says, "When I saw this old car in a goldenrod field I was seduced by its peeling chrome and chalky paint."[51]

The artist also has great appreciation for placing birds in a biologically appropriate context: "I needed a yellow bird, geographically correct, to be the focal point of the painting." And "the grosbeak added the perfect note for it seemed to engage in a 'dialogue' with the flowers, and its massive beak reminded me of the big fender of the Studebaker." She also has, of course, a great eye for finding and presenting aesthetic qualities. She approached the composition "in an abstract way, arranging colors and shapes in a long rectangular canvas . . . primed . . . with bright red before painting the picture." Why? "I'm curious to see if it will get a warm tint as it gets older, as oil paint becomes more transparent as it ages."[52]

Viewing the Science

Landscapes, even unusual ones like this, can convey the dynamic nature of natural contraction and expansion. In this case, the Studebaker Champion was losing more than paint; it was on its way into the history books and the scrap heap. In contrast, the range of the Evening Grosbeak was expanding, and perhaps this made the bird the new champion. Not only was the bird an excellent aesthetic choice in this painting, it made biological sense as well. In terms of territoriality, the behavior recorded here is accurate: birds may aggressively defend their territories against perceived intruders, sometimes even confronting reflections of themselves in shiny chrome surfaces. Some viewers might see the narrative as a provocative metaphor for the needs of birds to regain their hold on habitat overrun by the automobile. As Pépin notes, "'Champion' tends to remain in people's minds, and from the reactions and comments they make when they see it, I have a theory that it's from seeing a car, sacred and venerated in our society, in this state of rustiness."[53]

PLATE 56. *New Ark*, 1991, by Ray Harris Ching

PLATE 56 Photo-Realistic Style: Realistic but Not Real

The Narrative
In this painting each species is instantly recognizable, and each color, texture, and position is convincing despite the implausible prospect that these animals would ever gather together. The dubious assemblage thus invites a closer look. Now engaged, viewers may want to understand why such a quiet, calm scene seems stressful, why some eyes are open, and why the tractor feed margin of fanfold computer paper is included. No matter how each of us interprets the assemblage, the aesthetic elements of the painting—the crowd in the back and the empty space in the foreground, the beauty and paradoxical vitality of the animals who are dead asleep and whose privacy we have invaded—draw us ever further into the scene.

The New Zealand–born British artist, Ray Harris Ching, wrote the following caption for this painting: "They are all travelers [who I] made as comfortable as I could for the journey. . . . In the end I had a grouping of unrelated animals, in front of which lay a pair of beautiful Diana monkeys. Only the male is awake and therefore conscious of its surroundings (if not the circumstances of its predicament), and all of the other travelers are asleep—except the dove, who had its part to play in the first ark and perhaps anticipates being called on again.

"But all else is sleeping: the beautiful rooster whose tail folds under the arm of the monkey using the feathered back as a pillow, the Guineafowl is asleep, the plate-billed mountain toucan sleeps on the floor with its beak over the monkey's tail, the Avocet and squirrel are asleep and lie in the sprinkling of hay and straw." About the unnatural grouping, Ching says, "I felt the thing would work best only if each of the birds was so convincingly painted that you would have to believe they were here, together, and [I felt] that because they were so very beautiful, you would want to believe it."[54]

Viewing the Science
Computer paper dates this "new ark" in which the animal passengers adopt behavior uncharacteristic of groups seen in nature or in captivity. What has brought a toucan, dove, rooster, Guineafowl, Avocet, squirrel and two Diana Monkeys together, if not humanity? Just as the environment modifies behavior and just as humanity modifies the environment, humanity is modifying the behavior of these animals. Their response seems to be to wait it out, figuratively to sleep through the era of human assaults on the environment—assaults also causing the biogeographic expansion of numerous species that makes them dangerous invasive pests in their new neighborhoods. As the artist clarifies, "*New Ark* is an early painting in a body of work that has, by 2004, come to dominate my artistic output. For a decade or more, I have used images of birds to address concerns, sometimes specific, sometimes vague, about our impact on the natural world."[55]

Ching may initially seem an unlikely artist to represent this book's coverage of art about nature that has science as an orienting component, because much of his work—like this painting—addresses conservation or environmental issues metaphorically. But his extraordinary ability to bring the beauty and grace of these animals to viewers, and his way of imploring us to look both at the painting and at the impact of human behavior on these animals, is unrivaled. He writes, "The business of art is never to tell the viewer what they already knew." Ching's interest in delineating a biological context is as negligible as his pleasure in painting birds is palpable: "It is important to me to deny [the backgrounds]. I don't want to tell you anything about real nature studies. Other people can do that. I want to tell you about my concerns at this moment with this thing that I am painting."[56]

PLATE 57. *The Tyrants*, 1997, by Thomas Quinn

PLATE 57 **Painterly Style:
Real but Not Realism**

The Narrative
In this painting whimsical juxtaposition leads viewers to ask questions about an improbable relationship. The Great Horned Owl (*Bubo virginianus*) and the Rufous Hummingbird (*Selasphorus rufus*) are instantly recognizable, but why are they next to one another? The positioning, especially in a work entitled *The Tyrants,* prompts a closer look as we consider the degree of tolerance each bird has for the other. Does size matter here, or is the owl watching with studied indifference? Is the hummingbird an irritating presence, an avian mosquito, or is it resting under the owl's protection? Questions encourage us to meander through the painting's aesthetic elements—the richly colored precision fading into the faint, streaky blur.

The California artist, Thomas Quinn, gives us some guidance: "A Great Horned Owl awaits the gathering dusk and the time of dark supremacy. A tiny antagonist pauses momentarily, still owner of the day."[57]

Viewing the Science
Contrast and commonality link these two birds. One is a nocturnal predator, the other a diurnal nectivore; one is a hefty, silent flyer, the other a tiny, buzzing dart of a bird; one is probably monogamous, the other definitely a promiscuous male whose relationship with its young ended after copulating with their mother. Both of these birds are seen as tyrants within their separate realms, and their appearance together leads us to ask if the birds are out of position. Could this be an act of defensive commensalism, where the hummingbird is protected at no apparent cost to the owl? Would the hummingbird prefer the owl to leave, or vice versa?

This painting is quintessential Thomas Quinn: The viewer cannot help but focus on the face of the owl before drifting down the streaks to notice the hummingbird, and then return to the owl's face to assess the relationship of the two birds. Quinn captures the moment using pinpoint realism within a painterly rendering that unifies the composition. Producing such a composition involves a series of cautiously made technical decisions. Quinn says, "I tend to advance like a carpenter, with a plan and some virtuosity with tools. I hate to tear out something and redo it. I want to get it right the first time, every time."

Whereas Ching in Plate 56 employs photo-realism and suggests the idea of figurative rest and waiting, Quinn here employs a painterly style and conveys the idea of actual rest and waiting.

PLATE 58. *Long-billed Curlew,* Palo Alto Baylands, California, 2004, by Tom Grey

PLATE 58 **Medium: Photographs and Realism**

The Narrative

Photographs of moving birds can be taken from a stationary position with the aid of luck. Tom Grey took this shot of a Long-billed Curlew (*Numenius americanus*) using a high-quality intermediate length lens. The camera was hand held, and the photograph was taken without a flash in evening light. The photographer's patience in waiting for the right conditions and situation, his skill, and his quick, last-moment reactions show us the bird with clarity, although Grey credits the Image Stabilization technology in the camera for reducing the effect of camera shake. "I was walking back from the platform at [the Palo Alto] Baylands [preserve on San Francisco Bay]. This bird circled around and flew right across the walkway in the light of the setting sun against the blue sky and I got off several shots of him panning along. I give my little Digital Rebel a lot of credit for getting and holding focus. This camera gets a lot of grief, but it's given me lots of good shots."[58]

This particular photograph is of interest for another reason. It looks sensational online and sensational but often small in print. The small print size reflects heavy cropping from a 6-megapixel fullframe image; the cropping was required to bring the bird forward. Only new artwork based on this image could return the bird to a truly satisfying size when exhibited. As Grey explains, "The really good bird photographers would question anything cropped to more than about 75 percent of full frame. And this one is only 10 percent! But radical cropping is pretty much a necessity when shooting with only about 400 mm of reach, and not having the patience to sit in a blind or spend two hours crawling on my belly to get really close to a shorebird without spooking it."

Viewing the Science

This photograph of a Long-billed Curlew in flight encourages us to imagine the drag on its bill and the energy required to hold such a heavy and awkward body part in balance. This impression gives us reason to suspect that the energetics of flight may be one of the reasons these birds are short-distance migrants compared to their shorter-billed relatives. In fact, many nonbreeding Longbills spend the entire year in their winter habitat.[59] The photograph also encourages us to speculate about the habitat left behind and the curlew's activities prior to departure. The mud on the very long bill suggests a visit to the shoreline, and its height on the bill suggests deep foraging.

Even when a great photograph is available, a photo-realistic painting may be more effective in conveying a certain message because the artist can more precisely select and perfect the elements of the scene. Photography is essentially democratic: what the camera sees is what the photographer gets. It may thus be too democratic to highlight a key element. Even when the subject and the setting provide an opportunity for a great photograph, the photographer may be unable to correct problems with the lighting, the focus, the angle, or the degree of contrast. And even when a photographer optimally catches a subject on film or in bytes, the resulting image cannot always convey what a photo-realistic painting can. Sally M. Berner's photo-realistic painting of turkeys, for example, benefits from her decision to provide no more than an impression of the background, just as Vadim Gorbatov's painting of a lek under attack benefits from his ability to provide more than an impression of the high-speed chaos (see Plates 16 and 69).

PLATE 59. *Long-billed Curlew*, 1998, by Chris Bacon

PLATE 59   **Medium: Painting and Realism**

The Narrative

Is this a real curlew or a painted one, a real stretch of sand habitat or an imagined one? It is hard to tell. Yet anyone familiar with these birds recognizes this one instantly, and the anonymous setting seems credible because the habitat is appropriate.

In 2004 the Leigh Yawkey Woodson Art Museum named the Canadian artist Chris Bacon as its twenty-sixth Master Artist, and this curlew demonstrates why he deserved the honor. It also demonstrates Bacon's familiarity with birds in the wild. Bacon explains: "Birds, for me . . . are the embodiment of nature. . . . I've become more aware of the subtleties and complexities that surround and affect them. Birds as subjects are beautiful enough . . . the birds that occupy my paintings have also become vehicles to help me explore and understand some of those effects."[60] In the process, he helps viewers as well. This is archetypal Science Art.

Bacon provides a glimpse into his creative process. The painting of the curlew was "originally conceived as a moonlit scene," he says; "the idea was to have the viewer travel through the piece, stopping at strategically positioned 'bright spots' and high contrast areas before ultimately stumbling on the curlew. Quite often that is the only time we see birds on a beach at night, when they move because we get too close. Unfortunately, my early attempts did not live up to my expectations, but the composition and original concept was worth pursuing, presenting me with the opportunity to introduce other elements: for example, a menacing foreground shadow and disruption or chaos in the patterns of sand, creating an irritation in an otherwise serene setting. This introduction is unsettling, an uninvited 'presence' is established and reason enough for the curlew to consider moving on."

According to Bacon, on occasion—and invariably unintentionally—he is that uninvited presence. "Whenever you encounter wild birds, they will allow you to approach only so close before they move on. Sometimes when I'm out walking, I don't notice the little creatures until it is too late. I'm within that 'safe' limit only for a moment, and get just a brief look before the birds disappear. Such tense split-second encounters inspired this portrait of a Long-billed Curlew." Here we can see the risks and benefits of photography versus painting. The flight-prone bird subject might appear to give the advantage to the photographer: just one click captures it. But often the bird is gone in an instant—and then the painter has the advantage, because an image is stored in the painter's head. That moment of expectant flight can be rendered from memory, and particular highlights can be introduced to guide the viewer.

Viewing the Science

At 23 inches (59 centimeters) in length, this is North America's largest shorebird. It uses its very slender, very long (9-inch [23-centimeter]) bill to probe for burrow-dwelling prey that it cannot see and to nab a variety of other prey that it can see.[61] From a distance, even when the observer cannot distinguish the details of its bold cinnamon plumage, the bill prevents us from confusing the Long-billed Curlew (*Numenius americanus*) with a Whimbrel (*N. phacopus*) or other curlew species.

PLATE 60. *Eskimo Curlew*, 1957, by Charley Harper

PLATE 60 **Medium: Minimal Realism— When Less Is More**

The Narrative

This serigraph portrays an Eskimo Curlew (*Numenius borealis*) foraging on a bivalve. Even though the terrain is stark, the bird is well camouflaged. The image includes far fewer details than a photograph would, but each one counts, and the artist, the minimal realist Charley Harper, presents them to great advantage. Photographs can overwhelm viewers with details: they include everything in the frame. Sometimes photo-realistic art does, too, even though artists typically discard extraneous elements to showcase critical ones. If realism conveys as much as it can, minimal realism does the same using fewer elements.

Harper worked at excising irrelevancies for forty years, until his death in June 2007. By reducing his subjects and their ecological context to the most elementary visual terms he enhanced them. As he put it, "When I look at a wildlife or nature subject, I don't see feathers, fur, scapulars, or tail coverts . . . none of that. I see exciting shapes, color combinations, patterns, textures, fascinating behavior and endless possibilities for making interesting pictures." He would say that he "doesn't count the feathers of the birds he paints, just the wings."[62]

Harper was formally trained, but his interest in realism waned as he became convinced that sticking to it would "reveal nothing about the subject that nature had not done better," and that "the constraints of recreating the illusion of 3D on a 2D plane would only grow more burdensome." In his serigraphs, there is more than meets the eye, but the eye must wander through suggested shapes and designs to encounter the definitive aspects of the environment, the ecology, and the behavior.

Viewing the Science

Birds use their legs to help regulate their temperature, and when they are seen standing on only one of them, the air or water temperature is probably not warm enough. Before migrating, members of this formerly abundant species add a thick layer of fat that raises the birds' weight to what was formerly a marketable 1 pound (0.45 kilogram). They used to be called "doughbirds," and in Labrador, one of their migration stopover points, local residents used to kill and preserve them for winter food.[63] Eskimo Curlews, if any are left, are now so endangered that their breeding sites are secret or unknown.

These migrants to the far north sometimes find themselves in suboptimal Arctic conditions and may depend heavily on defrosted crowberries as a carbohydrate-rich stopgap until warming days bring a greater abundance of food. They have been known to forage on enough of the blue berries for the messy deep purple juice to stain their bills, legs, and plumage.[64]

ROOM 8

The Importance of Captions

The three paintings in this room demonstrate the value of writing captions that decode the underlying science.

As we have seen, works of art sometimes explain themselves, but sometimes more is needed. This is especially true of art that communicates about nature or science. On occasion, the ancient Egyptians included captions within the art itself. We saw this in the Lower Gallery: the statue of King Chefren is inscribed with hieroglyphics that specify the extent of his realm (see Plate 4), and the pelican painting on the wall of Horemheb's tomb even mimics a portion of the hieroglyphic text that it contains (see Plate 12). Captions still occur as an integral part of some art forms, but with the exception of cartoons, films, advertising, television, and digital media, they are rarely found within works of art today.

Captions accompanying paintings vary widely, but they typically give the title, the artist, and the date the work was completed. They might also contain other information, such as the subject of the painting and its location, whether it is part of a series or a study for a larger work, the medium and size. This is routine. The information keeps straight the historical record about the work. But for Science Art, captions play a central role, for expecting the viewer to take in the whole range of biological or ecological relationships suggested in the portrayal may be unrealistic. For audiences unfamiliar with biology, text is needed so they can decipher the biological content of the image. Exhibition catalogues have met this sort of need in the past. In the 1850s, for example, they helped audiences unfamiliar with the iconography of allegorical paintings understand what they were looking at.[65] Today publications featuring art about nature with a scientific focus ought to meet the corresponding need for biological explanations.

Although there is no formula for writing successful captions, in most venues captions for Science Art have some things in common. Exhibit specialists advise that less is more. Studies suggest that readers prefer twenty-five to seventy-five words, that with multiple images there is an advantage to having an overarching theme, and that it helps to strike a balance between informing viewers and urging them to interpret the art on their own.[66]

In public spaces and other venues where people are on the move, caption writers must satisfy the many who only glance at the art as they hurry past to see if a name or a title looks interesting or familiar, and the few with time to invest in reading about biology or ecological relationships.

In galleries and museums, if a work of art does not hold the viewers' attention, the caption may draw them in and lead them back to the art with heightened interest. Good captions cannot make middling art or middling exhibits great, but good captions can improve them. Catalogues and even inexpensive brochures help viewers approach the art, especially when exhibitors include descriptions of an organism's biology or ecological interactions.[67]

Online—where the number of seconds it takes to load an image helps determine whether a viewer will stick around to look at it—time is a factor, and limited text is a plus. Thumbnails (very small, low-resolution images that pop up quickly) with attached one-liners that link to larger images and explanatory text are the obvious preference.

Captioned art in periodicals, especially glossy magazines, is often provided to lure readers into the text, but it may also illuminate content that is difficult to either describe or visualize. Here, captions are usually descriptively spare, but readers are helped when a caption includes the name of the species portrayed, along with the Linnaean (latinized) name for international readers and researchers, the habitat type and its location, and the behavior portrayed plus its ecological or evolutionary significance. A more complete list, including nontraditional items, is compiled in Appendix 2.

The inclusion of full captions in catalogues may encourage emerging artists to develop the habit of composing a detailed caption for each image as it nears completion and to anticipate providing images for science publications as a measure of their professional reach.

PLATE 61. *Two Stories—Common Nighthawk*, 1994, by Carel Pieter Brest van Kempen

PLATE 61   **A Model Caption**

The Caption
This painting features the Common Nighthawk (*Chordeiles minor*), which breeds throughout North America from Canada to Panama, save for the extreme north and some southwestern deserts, and winters in South America. Common residents of many American cities, nighthawks spend their days sleeping on flat rooftops, where they also lay their eggs (usually two). They typically wait until sundown to take to the air, and their strange buzzing chirp blends in with their surroundings—resembling the sound of some electrical gizmo more than it does the call of a living creature. On summer evenings in Salt Lake City the artist, Carel Pieter Brest van Kempen, loves to watch them catch insects. Watching them, he is always struck by the contrast of two worlds: birds above and people below, each species seemingly oblivious to the other. Juxtaposing the two worlds in his fictitious scene required a bird's-eye-view.

Nighthawks are opportunistic birds who adapted well to urban sites and became common in cities after the mid-1800s, when gravel roofs were introduced. Where artificial light attracts moths, nighthawks often catch them—and once in a while they also catch the eye of those who, like this artist, are looking up from below.[68]

Additions When a Caption Is Used as a Catalogue Entry
Brest van Kempen brings a singular vitality to wildlife art. This may be due to his strong connection with his subjects. As he notes, "While many bird artists paint in order to share their experiences with the wider world, my motivation is more selfish. I usually paint as a way of seeing things I would like to see, but could never see in any other way. This includes hypothetical but plausible situations or impossible viewpoints. I paint in an extremely fussy style where every detail is more visible than it would be in a photograph. In the real world you can move in closer to a subject or change your position in order to discover more about what you're viewing. I try to design my work to give my viewer a similar experience." When painting this nighttime scene, he says, "I had great fun inventing the city details, incorporating lame jokes and inside barbs at friends and acquaintances. Close scrutiny even reveals a hint of my own face in the '67 Dodge van that served as my studio/home at the time I painted this."

Brest van Kempen's work has been exhibited widely across six continents, garnering numerous awards, including the Arts for the Parks Wildlife Award, three Society of Animal Artists Awards of Excellence, and Best of Show at the Pacific Rim Wildlife Art Show, the South Eastern Wildlife Expo, and many other major group shows. His representational style is ideally suited for the wide variety of birds he paints, including hornbills, bee-eaters, and members of the falcon family, particularly those found in the tropics. He has a presence on the Web, at www.cpbrestvankempen.com; www.artistsofnature.com; www.parcplace.org/PARCArtWeb/CarelPages/carel_pbv_kempen.htm; and http://king2.kingsnake.com/gallery/CPBVK, and in the printed archive; see especially his recent book, *Rigor Vitae: Life Unyielding; The Art of Carel Pieter Brest van Kempen*.[69]

PLATE 62. *Northern Mockingbird [Mocking Bird] (Plate 21)*, c. 1825, by John James Audubon

PLATE 63. *Brown Thrasher [Ferruginous Thrush] (Plate 116)*, 1829? by John James Audubon

PLATES 62 AND 63 **Writing Captions to Rectify Errors**

The Narrative

Like Audubon's other early paintings, these two portrayals of nest predation by snakes were produced as stand-alone compositions, which makes them easy to compare. (In later works, other artists produced the background.) Both show the harrowing attempts of the birds to defend their nests at the possible cost of their lives, but one is based on a misassumption. The error is difficult for many viewers to find. Indeed, in the absence of a caption, errors can go virtually unnoticed or even gain a life of their own, though not always to the detriment of the work or its value. In *Northern Mockingbird* (*Mimus polyglottus*), c. 1825 (Audubon's original title was *Mocking Bird*), it was not unreasonable to put a snake in the tree, for snakes do prey on eggs and young (Plate 62). Timber Rattlesnakes do not usually invade tree nest sites, however, and Audubon was so criticized for the portrayal during his lifetime that this plate is often referred to as one that he got wrong.[70] Nonetheless, the criticism served to bring attention both to this dramatic painting and to the similar but biologically correct *Brown Thrasher* (*Toxostoma refum*), 1829? (Audubon's original title was *Ferruginous Thrush*) —with significant financial impact (Plate 63).

Today, a Havell Edition print from Audubon's *Birds of America* of the correct painting, *Brown Thrasher*, sells for $15,000–$25,000, whereas *Northern Mockingbird* sells for twice as much: $30,000–$60,000.[71] Several factors explain the difference in price, but as with misprinted stamps coveted by philatelists, uniqueness in art prints inflates the price as well. The risk to artists in making a mistake or in portraying a research finding that later proves erroneous may be less than one might suspect—and may even work to the advantage of the artist or patron in the long run. Including natural behavioral relationships in art serves the patron, errors or not, and in the very long run errors can also have scientific value. Bird images found throughout the 30,000-year pictorial archive have left a record of what our forebears understood about birds, about nature, and about birds' place in nature. Noticing and understanding what they noticed and understood is instructive.

The Rectification

Errors in research results are overturned by subsequent studies. This self-correcting mechanism is one of the strengths of science. Occasionally art portrays the earlier results, or it portrays an aspect of nature inaccurately, or it fails to discourage viewers from inaccurately interpreting it. Captions provide an opportunity to set things straight.

As we saw with the birds in the monk's drawing book (Plate 20), errors can be replicated when art is copied, and sometimes imitation produces a succession of interesting mistakes. The former director of Sotheby's, T. H. Clarke, traced images of the Indian Rhinoceros (*Rhinoceros univornis*) from Albrecht Dürer's woodcut in 1515 to George Stubbs's painting in 1799. Dürer's rendering of the armor-plated rhinoceros suggests that he had never seen the animal; the extra horn extruding from the shoulder and pointing forward lends credence to this supposition. But he knew quite a bit about metalwork. He actually designed armor; his drawing of a visor for a jousting helmet is also dated 1515.[72]

Dürer's "plated" rhino was a frequent target of imitators and a model for copyists, as was his painting of a wing (*Wing of a Blue Roller* [*Coracias garrulous*], 1512), which the German artist Hans Hoffmann copied several times, adding the "AD" monogram, and which Theodor Josef Ethofer also copied, adding the occasional extra primary feather. Interest in Hoffman's wing continues. One of his copies was featured in a special Masters Exhibit at the National Gallery in Washington, DC, in 2006.[73]

Another much-copied work is an etching of a Great Auk (*Pinguinus impennis*) by Olaf Worm. The original was used as the frontispiece of the catalogue of specimens contained in Worm's Copenhagen Museum: *Museum Wormianum* (1655). The etching was a portrait of Worm's pet Great Auk, which is shown wearing a white collar. Later artists copying the frontispiece and unaware of the auk's solid black neck plumage included a white ring as prominent as the bird's vestigial wings.[74]

ROOM 9

From Real Public Venues to Virtual Ones

The two images in this room enable us to explore the use of public spaces—especially those in academic settings—to exhibit art about nature that has a scientific focus, and to consider the use of the Internet to post additional information about installations.

Living artists have few outlets for their work. Gallery space is limited, and much museum space is reserved for deceased artists. Fortunately, the use of empty walls in public spaces remains an underutilized boon. Certain locations are far better than others—and competition for their use may already be significant. Many well-positioned bare walls have yet to be converted into exhibit venues, however, and their conversion may meet initial resistance, so it might help artists producing informative images to seek bare walls in institutions of learning. There the audience is preselected to value it, and the chances are good that the wall owners are concerned about nature and conservation, predisposed to support the arts, and interested in promoting alternative teaching methods. Universities, high schools, public libraries, and corporate headquarters, among other such venues, often have an array of exhibit options in addition to miles of untenanted walls, where Science Art, free for the viewing, can attract a broad cross-section of passersby.

Like most universities, Stanford has many public spaces, some of which have been converted for use by exhibitors. For more than fifteen years a large, well-situated pair of display cases in the Falconer Biology Library has been available periodically for exhibiting Science Art. At Stanford, as at many other institutions, walls not designated as exhibit spaces might also be used for display, and the people responsible for managing them may welcome new installations. Artists approaching these managers with samples of their work might find it helpful to do so in the company of an advocate with official ties to the institution. The Stanford Faculty Club turned out to be a place in which the authors successfully negotiated the conversion of a bare wall into an exhibit area that now accommodates seven large bird pictures, captions, and a brochure holder. By posting a thumbnail of each image on a university-sponsored Web site, the authors have been able to direct new students and visitors to the exhibit. The Web-site presence later helped the authors arrange the installation of paintings on several other bare campus walls. The two pictures in this room are examples of the artwork offered for public view at the university.[75]

From space on bare walls to space on the Internet, viewers have many opportunities to stumble onto examples of Science Art. Narrowing the search, so that luck is not the operative factor, poses some difficulties. For exhibited images, public announcements specifying Science Art in the local media are key. For posted images, overwhelming search results remain a problem. In April 2007 a Google search for "Science Art" yielded an unwieldy 1,040,000 results, up from 395,000 in January 2005—further evidence that refined search criteria and Science Art databases are needed (see Appendix 2).[76]

PLATE 64. *Vultures and Crystals* [Oriental White-backed Vultures (*Gyps bengalensis*)], 2004/2007, by Darryl Wheye

PLATE 64 Short-Term Exhibits for Academic Use

The Exhibit

A five-month exhibit at Stanford University's Falconer Biology Library (November 2004–April 2005) entitled Changes in Conservation Status called attention to the enormous number of vulture deaths in India that occurred at the end of the millennium.

Since the early 1990s hundreds of thousands of healthy-looking vultures in India had dropped dead. The cause of death appeared to be scavenged food contaminated with diclofenac, a nonsteroidal anti-inflammatory drug that is given to livestock and tends to concentrate in their liver and kidneys. Autopsies showed that the internal organs of 85 percent of the dead vultures had a buildup of uric acid crystals, usually a telltale sign of kidney failure. The same birds tested positive for diclofenac. This was the first record linking wildlife losses to a veterinary drug. Those losses were significant: in the twelve years between 1992 and 2004, Oriental White-backed Vultures (*Gyps bengalensis*; shown here) had declined by more than 99 percent, and Long-billed Vultures (*G. indicus*) by 97 percent. After a slow start, steps were taken to phase out the drug and establish a captive breeding program to rebuild the vulture populations, some of which were plummeting by half yearly. The effects of diclofenac were later reported in Pakistan as well.[77]

Years ago Kenneth Brower warned, "When the *vultures* watching your civilization begin dropping dead . . . it is time to pause and wonder." The pause-and-wonder phase may have lasted too long in the case of these birds, for the repercussions are widespread and troubling. For two millennia the Parsi have laid out their dead at the top of Towers of Silence to be quickly eaten by vultures, but few vultures show up now. Elsewhere, especially at dumps, rat and feral dog populations climbed sharply in the absence of vultures. These second-rate scavengers are less efficient, and they harbor rabies and other diseases. Even the deaths of sacred cows now pose problems that, in some places, have elevated into health hazards.[78]

The Venue

Science Art is not yet commonly exhibited. But it could be, and images like this need not be exhibited in a biology library to gain attention. Teaching is often compartmentalized; learning is not. Formal education has provided conditions favoring the development of discipline-specific departments and the evolution of curricula based on those departments. Thus it happens that science is taught in science class, whereas art is taught in art class. Because Science Art is relevant to both, is educationally useful, and offers an important way of drawing people's attention, it could create a productive zone of overlap between these two disciplines. Students, instructors, and even the public at large could find out about emerging environmental issues, for example, or their resolution, in the zone of overlap. Science Art does not always have a specific agenda, nor should it, but it often provides a medium for environmental education that reaches beyond particular issues (see Plate 65). The wide appeal of Science Art is reminiscent, in some respects, of the large number of nature images produced in the 1600s, when the Golden Age of Dutch art benefited from Dutch dominance in commerce, which fostered the exploration of remote regions—leading in turn to the production of images featuring the natural history and species from these places.[79]

PLATE 65. *Avian Engineering* [Bushtits (*Psaltriparus minimus*)], 2005, by Darryl Wheye

PLATE 65 **Long-Term Installations for Academic Use**

The Exhibit

This painting, which is installed at Stanford's School of Engineering, calls attention to biomimetics, the practice of basing designs on models found in nature. The themes of the Science and Engineering Quad at Stanford University, as described in its Web site, are connecting and gathering: The quad connects formerly dispersed engineering departments, and the walkways, tunnels, plazas, and gardens connect its collection of buildings. Within those buildings faculty and students gather and mingle, and outside the buildings pedestrians gather or circulate among carefully spaced trees. In that milieu, structural reminders of nature can be assets, especially to engineers and scientists, where a keen eye might notice spider webs or nests or other constructions in the surrounding vegetation.

The painting was produced to encourage a closer look at the housing that birds construct. Its life-size Bushtit (*Psaltriparus minimus*) nest is suspended from the branches of a Locust (*Robinia* sp.) overlooking the Packard Electrical Engineering Building and the Gates Building beyond. The cutaway exposes grid lines like those seen in technical drawings and invites consideration of avian engineering. Bushtits use spiderwebs to bind lichen, moss, string, fiber, plant down, and flowers, along with the occasional twig, cocoon, or feather, into a durable artificial tree hole that is as elastic as spandex and as waterproof as Gore-Tex.

The construction of the School of Engineering Quad took $140 million, legions of workers, and decades of effort. The construction of the Bushtit nest took 140 million years of evolution, a pair of 4-inch (10-centimeter) birds, and seven weeks of work.

Humanity has successfully co-opted the construction plans of any number of species; biomimetic successes are all around us.

The Venue

Science Art is no more commonplace in long-term installations on academic campuses than it is in short-term exhibits. For now, however, curricula in higher education do not foster its production. Students cannot currently find art courses teaching Science Art, and that is unlikely to change overnight. Interested students can certainly urge professors to develop such courses, but adding new subjects to courses of study takes time. Nowadays if students encounter Science Art, it is likely to be inserted into ongoing coursework. Adding Science Art to environmental science and other biology courses and to studio art and art history courses would help raise its visibility and widen the path for those wishing to produce it. Seemingly unrelated fields may forge unexpected connections as a result: engineering classes could take up biomimetics—discussing, for example, how the rigid wing flaps of aircraft resemble as much as possible the curvature of bird wings during landing.[80] Images showing how movement against water resistance helped to shape penguins, loons, and diving ducks—as well as submarines—can provide convincing and memorable arguments.

Yet given the press of faculty life, few professors have time to seek out and utilize Science Art in a way that may engage students. Two steps that could be taken to promote the wider use of Science Art—universal, fee-free permission for the educational use of images, and the production of a comprehensive Science Art database to help locate images, the artists producing them, the galleries and museums exhibiting them, and the books and periodicals publishing them—could, taken together, make a significant difference.[81]

ROOM 10

Science Art, Birds, and Perceptions of Nature

The four images in this room summarize the themes of this book. The first image, a rather stark impressionist painting of an urban woman by Édouard Manet, prompts us to think about captivity, both ours and that of our pet birds. It also allows us to explore the adoption of the term "Science Art" to describe the kind of art featured in this book. The second image, which was published with the report of the Ivory-billed Woodpecker's apparent rediscovery in 2005, and the remaining two images—paintings of Black Grouse by different artists from different centuries—allow us to explore how Science Art can heighten our perceptions of nature as exemplified by our strong and growing interest in noticing and watching birds.

Science Art

As we saw in the Lower Gallery, the connections among science, art, and nature extend back to the Paleolithic. The archive of art showing those connections is enormous. Even a cursory assessment of public venues and the opportunities for new ones suggests that it should not be difficult to raise the visibility of this kind of art and benefit from its increased use in public spaces, on the Internet, in the various print and electronic media, and in teaching.

The category Science Art is not recognized as a subject heading in libraries, reference books, or databases, including bibliographies and art and image databases. It is not recognized by arts commissions, granting agencies, many art groups, major nature-related art competitions and exhibitions, or galleries and museums.[82] The main hurdle in increasing the use of this kind of art in teaching is name recognition. If, for example, Science Art becomes a widely recognized label, parents will more readily find books to introduce their young children to science, art, nature, and environmental protection. Many books already exist.[83]

Although Science Art is hard to find, teachers already have a useful online resource, Testbank, sponsored by the Environmental Literacy Council, which provides them with tools to help students become environmentally literate and acquire the analytical skills necessary to evaluate scientific evidence. Students lacking these skills have difficulty differentiating those policy choices that are sound from those that are not. The Environmental Science Testbank itself could serve as a model for teaching students how to evaluate and produce Science Art and how to interpret and write Science Art captions.[84]

Professors can breathe life into a lecture on the rise of ornithology as a science through the use of evocative but rarely seen historical images. As we also saw in the Lower Gallery, paintings document the transition of birds from objects of interest to subjects of organized study. The timeline in Appendix 1 lists quite a few of the transitional events and the artists who produced related bird images.

Interdisciplinary courses in environmental science and environmental studies have become common in colleges and secondary schools, providing a large audience with both an interest in and a growing commitment to nature and to the local environment, and providing as well a new and receptive audience for artists. Here, too, the development of large graphics databases making artists and their work more readily available will make a difference.

Many studio art programs already encourage collaboration and connections with scientists and other scholars. More universities and high schools are expanding their focus on the arts to include collaborative interdisciplinary programs that are helping to integrate the arts into their educational mission. In 2006, for example, the Stanford Institute for Creativity and the Arts (SICA) launched Creative Risks, "a series that brings in contemporary artists to collaborate with students in multiple areas of the university." A wealth of art featuring birds and nature is available to teachers and students, especially images produced during the rise and establishment of the Scientific Revolution (c. 1500s–1700s), the Age of Naturalism (c. 1830–1874), and the Environmental Movement (1960s on), as are books.[85] As interest in the environment has fostered interest in nature and nature-related science, and vice versa, Science Art has become central in connecting us to all three.

Birds and Perceptions of Nature

Why do we enjoy nature, and why does Science Art succeed in connecting us with it? Some of us, we suspect, have what E. O. Wilson has called "biophilia"—that is, we feel entangled with the web of life, recognizing a kinship with the other living creatures with which we share the planet. Some of us enjoy the outdoors and take some pleasure in the order of nature—perhaps instinctively recognizing the importance of relationships within ecosystems and understanding that this forest, this understory, and these native birds somehow belong together.

Why, in particular, are so many of us fascinated with birds? First and foremost, they get our attention. Evolution has not only shaped the relationship of birds and other organisms to the rest of the biosphere, it has also selected the perceptual filters with which we interpret birds and everything else we see in nature. Humans are among that tiny minority of mammals that possess color vision, and we have perceptual systems that are unusually attentive to fast movements. We are also a curious species, interested in observing other animals and often finding analogies with our own behaviors.

Indeed, we have created—as did our ancestors—a virtual anthology of iconographic animals, including birds, some of which we have endowed with humanlike qualities. The portfolio of anthropomorphized creatures includes semirealistic ones (Bambi) and caricatures (Wile E. Coyote). Woody Woodpecker is a bit of a caricature, but his woodpecker behaviors have made him an attractive icon in the cartoon world (he even has a star on Hollywood's Walk of Fame). In the real world we still make icons out of living birds, however rarely.

Native Americans and the early naturalist-explorers of the southeastern United States endowed its largest woodpecker with iconic status. The Ivory-billed Woodpecker's spectacular plumage and bill and its unique foraging behavior—it stripped large segments of bark from trees to gain access to the larvae beneath—drew admiring attention in the nineteenth century. The origin of its local moniker, the Lord God Bird, is told in this passage from the recent avian biography by Phillip Hoose: "The most telling nickname of all came from an expression of awe, an exclamation uttered by those who suddenly caught sight of an arrow-like form ripping through the highest leaves of a deep forest, unfolding its three-foot-wide wings to the size of a flag, and then finally swooping straight up to sink its mighty claws into the thick trunk of a cypress tree. At such moments, sometimes all a dumbstruck witness could say was 'Lord God, what a bird!'"[86]

By the 1930s the Ivorybill's abundance had so dwindled that it was featured in *Natural History* and *National Geographic* articles. One of the authors remembers his fascination with Arthur Allen's Cornell University expedition to Louisiana to photograph and record the vocalizations of Ivorybills in 1935. The Lord God Bird had become a premonitory symbol of what environmental scientists call the "extinction crisis." It was 1944 when Don Eckelberry, employed by the National Audubon Society, visited the newly logged Singer Tract in northern Louisiana and became the last North American artist to paint the Ivory-billed Woodpecker in the wild. Eckelberry left us with remarkable pictures—in words as well as watercolors. Here he records his first sighting of what was then the last known female: "She came trumpeting in to the roost, her big wings cleaving the air in strong, direct flight, and she alighted with one magnificent upward swoop. Looking about wildly with her hysterical pale eyes, tossing her head from side to side, her black crest erect to the point of leaning forward, she hitched up the tree at a gallop, trumpeting all the way. Near the top she became suddenly quiet and began preening herself. With a few disordered feathers properly and vigorously rearranged, she gave her distinctive double rap, the second blow following so closely on the first that it was almost like an echo—an astonishingly loud, hollow, drumlike *Bam-am!* Then she hitched down the tree and sidled around to the roost hole, looked in, looked around, hitched down beneath the entrance, double-rapped, and went in."[87]

Audubon and other traveling naturalists were also enchanted by the Ivorybill, and Audubon's painting of the bird is one of the most frequently reproduced Ivorybill images. How did Audubon manage to make his Ivorybills look so animated? Birds are apt to be skittish and reluctant to pose. Field glasses were rare and their optics were crude, and there were virtually no cameras, so artists often had to base their renderings not on living specimens but on birds shot expressly for the purpose. Artists were not alone in their use of the gun when doing detailed work. Ornithologists liked firm evidence for identification, and an old saying from the early days of competitive birding was "The only sighting I'd trust is over the barrel of a 20–20 shotgun." Collectors and artists, operating in a time when the bounty of bird life seemed inexhaustible, employed guns freely, probably too freely. The artists, however, realized that a more lifelike subject was needed. Audubon noted his dissatisfaction with museum-like presentation and its lifelessness: "The first collection of drawings I made of this sort were from European specimens, procured by my father or myself. . . . They were all represented strictly ornithologically, which means neither more nor less than in stiff, unmeaning profiles." Audubon sometimes got around this difficulty by threading wires through fresh specimens to position them as if alive. Positioned models were tremendous assets. Even artists not specializing in birds, including the British portrait painter John Singer Sargent (1856–1925), created supports for the birds included in their compositions.[88] For many, a wounded bird might be preferable to a limp dead one, although that, too, posed problems.

Other ornithologically minded artists made efforts to find and paint the Ivorybill, but those of Alexander Wilson were surely the most determined. Wilson's painting appears in *American Ornithology,* the first scientifically sound book on birds with colored plates published in the United States. It appeared as a series of volumes issued between 1801 and 1814. Wilson's account of his adventures with his subject provides a different kind of confirmation of the value of a live model—and of the vigor with which the Ivorybill could attack bark.

"The first place I observed this bird at, when on my way to the south, was about twelve miles north of Wilmington in North Carolina. There I found the bird from which the drawing of the figure in the plate was taken. This bird was only wounded slightly in the wing, and, on being caught, uttered a loudly reiterated, and most piteous note, exactly resembling the violent crying of a young child; which terrified my horse so, as nearly to have cost me my life. It was distressing to hear it. I carried it with me in the chair, under cover, to Wilmington. In passing through the streets, its affecting cries surprised every one within hearing, particularly the females, who hurried to the doors and windows with looks of alarm and anxiety. I drove on, and, on arriving at the piazza of the hotel, where I intended to put up, the landlord came forward, and a number of other persons who happened to be there, all equally alarmed at what they heard; this was greatly increased by my asking, whether he could furnish me with accommodations for myself and my baby. The man looked blank and foolish, while the others stared with still greater astonishment. After diverting myself for a minute or two at their expense, I drew my Woodpecker from under the cover, and a general laugh took place. I took him up stairs and locked him up in my room, while I went to see my horse taken care of. In less than an hour I returned, and, on opening the door, he set up the same distressing shout, which now appeared to proceed from grief that he had been discovered in his attempts at escape. He had mounted along the side of the window, nearly as high as the ceiling, a little below which he had begun to break through. The bed was covered with large pieces of plaster; the lath was exposed for at least fifteen inches square, and a hole, large enough to admit the fist, opened to the weather-boards; so that, in less than another hour he would certainly have succeeded in making his way through. I now tied a string round his leg, and, fastening it to the table, again left him. I wished to preserve his life, and had gone off in search of suitable food for him. As I re-ascended the stairs, I heard him again hard at work, and on entering had the mortification to perceive that he had almost entirely ruined the mahogany table to which he was fastened, and on which he had wreaked his whole vengeance. While engaged in taking the drawing, he cut me severely in several places, and, on the whole, displayed such a noble and unconquerable spirit, that I was frequently tempted to restore him to his native woods. He lived with me nearly three days, but refused all sustenance, and I witnessed his death with regret."[89]

The Ivorybill has what an ecologist would call extreme niche specialization. Wilson's captive showed an impressive capacity for hammering wood, but in the wild the way the birds attack trees is very different from that—and from the tactics of other woodpeckers. Eckelberry captures the style in this passage: "She attacked the bark by rearing back, her head usually off to one side, and striking at an oblique angle. Often the loosened bark was flipped off, after one or two blows, by a quick movement which appeared to combine a slight twist and a lateral flick of the head. Hammering was not rapid or persistent, but the bird's long neck gave tremendous thrust, for when fully applied these glancing stabs knocked free pieces as much as half a foot long and 3 or 4 inches across."[90]

How does this give the Ivorybill a competitive advantage? By peeling bark, they are able to eat the grubs—especially the meaty Longhorned Beetle (Cerambycidae) larvae that are thick as a finger and long as a palm—hidden beneath an entire segment of the bark, prey that other woodpeckers would have to capture one at a time by well-targeted pecking. James Tanner, the University of Tennessee biologist who studied the birds more intensively and for longer than anyone else, made mention of "this private stash of grubs." At the end of his career, he gave the retirement lecture that was a University of Tennessee tradition. After describing his studies, he had to describe the eventual deforestation of the Singer Tract and was too overcome with sadness to continue his talk to the end.

As recently as 1987, there were generally accepted reports of another subspecies of Ivorybills in Cuba, so the possibility of their continued existence is surely possible. For the U.S. subspecies, there was no verified sighting of the Ivory-billed Woodpecker after 1944, but news of its possible existence in the Southeast's riverbottom forests circulated frequently on the basis of unconfirmed (but passionately claimed) individual sightings. No scientifically credible claim came forth until 2005, when a group of ornithologists from Cornell University published, in *Science,* an account supported by repeated sightings by experienced observers, a video, and measurements taken from frames of the video. The paper passed a review by other specialists and attracted international media attention, but the bird has not been verifiably sighted in the seasons of searching since then.

Other specialists regarded the report with skepticism, and ornithologists of substantial reputation offered alternative interpretations of the evidence. Birders take pride in their expertise, and any field trip with a group of experts can generate its share of tense disagreements. The scientific literature is filled with claims and counterclaims. One knowledgeable participant in the debate is David Sibley, the distinguished ornithologist and artist known to contemporary American birders as the author and illustrator of the deservedly popular series of field guides published by the National Audubon Society. He was not convinced by the evidence that what the Cornell team saw and filmed was an Ivory-billed Woodpecker and not a Pileated Woodpecker. We might not know for some time who was right. If no new and convincing evidence appears within the next few years, most scientists will probably conclude that the iconic status of the bird has led to yet another triumph of hope over experience. In its own News section, *Science* mourns that likely loss with a story entitled "Gambling on a Ghost Bird."[91]

Has the Lord God Bird survived despite the environmental challenge it has faced? We do not know. Even a large, famous, very rare bird can be hard to find in a big territory, even by watchers with good technology and a lot of determination. For some, hope remains. Some searchers even now are out looking in one bayou or another like the one William Faulkner describes: "He stood against a big gum tree beside a little bayou whose black still water crept without motion out of a cane-brake, across a small clearing and into the cane again, where, invisible, a bird, the big woodpecker called Lord-to-God by negroes, clattered at a dead trunk. It was a stand like any other stand, dissimilar only in incidentals to the one where he had stood each morning for two weeks; a territory new to him yet no less familiar than that other one which after two weeks he had come to believe he knew a little—the same solitude, the same loneliness through which frail and timorous man had merely passed without altering it, leaving no mark nor scar, which looked exactly as it must have looked when the first ancestor of Sam Father's Chickasaw predecessors crept into it and looked about."[92]

Now, sixty years after Faulkner wrote these words, the timorous—as well as the bold—who seek nature without the ambition of altering it may trek into the bottomland and pause to listen for a woodpecker by a still bayou. Or they may stride through a meadow, mindful of a Song Sparrow clinging to a bramble. Others of us might snatch the odd moment to take a virtual hike by viewing—entering—an image, like one of the four in this room. Entering the first image, we are reminded why we seek both refinement and nature; by entering the second, why we care so much about the rediscovery of a species no one had seen in two generations; and by entering the third and the fourth, why an aggregation of birds tells us a little more about our place in the world.

PLATE 66. *Young Lady in 1866*, 1866, by Édouard Manet

PLATE 66 **Interest in Birds: Viewing Nature through the Lens of Culture**

The Narrative

Édouard Manet (1832–1883) was well known for sidestepping narrative in his paintings. Nonetheless, this picture tells a story. We include it here because this narrative of culture provides a contrast to the narratives of nature used throughout this book. In 1866 both Manet and his friend Gustave Courbet painted portraits of women and African Grey Parrots (*Psittacus erithacus*); Manet's is shown here. Edgar Degas, another friend, portrayed a woman in a dressing gown accompanied by a parrot. Was it coincidence, or was the inclusion of exotic birds in portraits not particularly unusual?

The popularity of exotic birds was high. The bird artist Joseph Smit produced 100 images for Philip Lutley Sclater's *Exotic Ornithology,* the first volume of which was published the same year Manet painted his woman and parrot, 1866, and by 1869, Otto Finch, a specialist in exotic ornithology, and Gustav Hartlaub, the preeminent German ornithologist of the day, had published a two-volume book on parrots. The portrayal of pet birds is a tradition with a long history. They are found in the art of Greece and Rome and in the highly popular works of Carel Fabritius, Peter Paul Rubens, Francisco Goya, and Sir Joshua Reynolds. The public, then, would have been familiar with the genre.

Today Manet's painting makes a statement about captives and our penchant for bringing icons of nature into our homes. The woman and the bird, both exotics and both seemingly captives, appear in a room so dark and silent that their vitality is startling. It is the unseen narrative of nature—the costs associated with removing the parrot from its natural habitat to live in this stark, dark room—that comes to mind. In Manet's day, parrots were prized for their intelligence and ability to speak, with the African Grey valued as the most fluent. They were seen as confidants, and the bird in this painting would have been presumed to know why the woman (Manet's model, Victorine Meurend) is holding the cord of a man's monocle in one hand and violet blossoms in the other and whether she is preparing for a formal event or has just returned from one. The parrot, portrayed as a companion, is found in Dutch and Flemish paintings of women; perhaps Manet was familiar with them.

The Role of Science Art

In 1874 the word "impressionism" was coined by a critic commenting unkindly on *Impression: Sunrise,* a painting by Claude Monet (1840–1926). The name stuck, and the sunny, flickering style that Monet employed gave representational art a leg up after photography, invented fifty years earlier, challenged its dominance. Monet was not alone in those early days. Manet joined in the movement by rejecting the age-old idea of the canvas as a window to look through and instead treated it as a paint-coated surface whose color and brush strokes were more real than what they stood for. Yet Manet never applied the term "impressionism" to his work.[93] Are impressionism and Science Art alike in that way—that is, are artists producing it without quite expressing with a label the fact that they are doing something different from the norm? It is not hard to do something without giving it a name—consider the man who was surprised to learn that all his life he had been speaking in prose.

Science Art, unlike impressionism, is not a style. It is a classification of content that cuts across all styles. If it has never owned its own long-lasting, inclusive, categorical label, is one really warranted now? The authors have been trying to answer this question for a decade, and we conclude that the term neatly describes a merging of science and art that weakens neither and aids both. Adding the phrase "that portrays nature" would tighten the focus and tell the viewer something about the relationship of the subject to its environment—but would be cumbersome. In our search we verified the presence of this kind of imagery throughout the pictorial archive, became familiar with contemporary bird art, and determined that there *is* a need for a label and that "Science Art" works.

PLATE 67. *Ivory-billed Woodpecker*, c. 1935, by George Miksch Sutton, on the cover of *Science* 308, no. 5727 (June 3, 2005)

PLATE 67 **Interest in Bird Conservation**

The Narrative

This watercolor by George Miksch Sutton, painted seventy years ago, was published in 2005. Here is the caption that *Science* magazine supplied: "A male ivory-billed woodpecker (*Campephilus principalis*), sketched from life in the Singer Tract of northeastern Louisiana in 1935. Long suspected to be extinct in North America after the tract was logged in the early 1940s, this species may have been rediscovered in the 'Big Woods' region of Arkansas, about [190 miles] 300 km north of the tract." (For more about the rediscovery of the woodpecker, see the introduction to Room 10.)

The Role of Science Art

When the editors of *Science* were deciding how to treat the surprising announcement, they quickly determined that the Ivorybill deserved to appear on the cover of the magazine. That left the question of what image to use. The choices were old photographs, skins of Ivorybills in museums, old Indian relics—any or all of which could have been used to represent the bird—but Sutton's painting was chosen. Why? Sutton was curator of birds at Cornell University and had joined Arthur Allen's 1935 expedition to photograph the Ivorybill and record its voice. Sutton was one of the only artists to see the bird in its habitat, and records indicate that he was thrilled. The expedition had arrived before mosquito season, and Sutton had "splashed behind [Arthur Allen, James Turner, and J. J. Kuhl] in a crisply pressed shirt and a well-knotted tie." Upon seeing the bird for the first time, Sutton wrote: "He called loudly, preened himself, shook out his plumage, rapped defiantly, then hitched down the trunk to look at me more closely. As I beheld his scarlet crest and white shoulder straps I felt that I had never seen a more strikingly handsome bird."[94]

When the paper reporting the 2004 observation was published, it received the attention of the U.S. Department of the Interior and the Nature Conservancy, leading to major financial commitments and land use decisions to protect the woodpecker's habitat. The bird's iconographic status clearly played a role in those decisions. Expanding protection to this new area offered hope that the bird, elusive since its reported rediscovery, might survive. As the biologist Jerome Jackson points out, "One of the most encouraging aspects of my searches for ivory-billed woodpeckers is that many prime areas are already under some sort of protection.... Can we save the ivory-bill? By as early as 1945, [James] Tanner did not think so. But if asked if humans might walk on the moon in twenty years, a scientist might then have said, 'It's not possible.' Humankind can do great things. I want to believe. If there is habitat, there is hope. If there are ivory-bills out there, there is hope."[95]

Preserving a species and its requisite habitat requires public commitment. When that commitment is weakened, we rely on images and news from researchers to reinvigorate it. In the case of the Ivorybill, we have a few relics that go back a thousand years, photographs from the first half of the past century taken by Arthur Allen and James Tanner, paintings by artists like John Abbot, Rex Brasher, George Miksch Sutton, and Don Eckelberry, who actually saw the birds, and paintings by Thomas Bennett, Carl Brenders, Guy Coheleach, and Julie Zickefoose, as well as many other contemporary artists, who were not in a position to paint the bird from life, all of which are available on the Internet.[96]

The role of Science Art—art plus captions—in decoding nature and enhancing our interest in its protection is seen in Sutton's *Ivory-billed Woodpecker,* his painting of a bird whose reported rediscovery caught the imagination of people around the world like no other conservation story of 2005. A bird species in great jeopardy can usually grab the attention of those who remain connected to nature, but this particular woodpecker caught the attention of everyone else, too, and it was their enthusiasm that lent tacit support to the carefully orchestrated protective efforts associated with the apparent rediscovery.

PLATE 68. *Blackcocks in Springtime (Orrspel)*, 1675, by David Klöcker Ehrenstrahl

PLATE 69. *Attack, Out of the Mist*, 1995, by Vadim Gorbatov

PLATES 68 AND 69   **Perceiving an Annual Ritual**

The Narrative

Black Grouse (*Tetrao tetrix*), also called blackcocks, appear here in paintings by two artists. In one, David Klöcker Ehrenstrahl portrays a habitat in which he could observe the birds doing something remarkable (Plate 68). The grouse have moved out from their sheltered roosting spots in the woodland to their traditional mating grounds, or lek, and formed a mating aggregation in the open where the assembled birds could easily see one another and where Ehrenstrahl could easily see them. Ehrenstrahl records the fervor of the lek as it might be captured on film, but each element of his painting remains in focus and can be studied separately. Vadim Gorbatov also records a Black Grouse lek, but he ratchets up the fervor, showing all but the thoroughly engaged central contestants fleeing the swooping approach of a Northern Goshawk (*Accipiter gentilis*; Plate 69).

The Role of Science Art

Although Ehrenstrahl painted his lek more than 300 years ago, we could conceivably see one in the same spot today, for every spring Black Grouse return to the same sites to form mating aggregations. Up to 200 males, but usually groups of 10–40, hold territories at a single, noisy, frantic lek, raising their lyre-shaped tails, spreading their white undertail coverts, hissing, and performing mock fights to hold or gain ground. They vie for more centrally located positions because centrality increases their chances for attracting a mate. Skirmishing is less over a specific female than over turf. Female participation is minimal. A gray hen solicits by crouching, the pair bonds for a few minutes, and then she leaves the lek. She will rear her young alone.

Gorbatov painted his lek in 1995. Like Ehrenstrahl, he conveys action and intensity, but Gorbatov's predatory goshawk and fleeing grouse look more realistic, even though Ehrenstrahl's grouse are anatomically accurate. Striking predators and fleeing prey act and react too quickly for us to observe them clearly, but the camera's eye, with its ability to stop the action, reveals what the human eye fails to detect and enables viewers to see acts like attempted predation that succeed or fail in a flash. The truth of the camera can also prepare artists with the equivalent of studies, preliminary compositions showing how a raptor swooping down for a kill looks and how birds recognizing the danger react. Perhaps this is a reason why early scenes of interacting birds, even when produced by an exemplary artist like Ehrenstrahl, appear slightly stiff today compared to those produced since the invention of photography.

Both artists portray an edge—a discontinuity between neighboring habitats, a place where habitat peripheries overlap. The area of overlap favors individuals found in both habitats, so it often contains more than the usual number of species, and often more than the usual amount of activity—making them ideal spots for naturalists of all sorts to visit. Neither artist takes us inside the area of overlap, but a well-crafted caption enables us to take a virtual visit.

If we were lucky enough to see a lek and watch an ordinarily peaceful landscape roil into activity, the chances are good that the sight would long be remembered. The sudden, unexpected burst of motion of even a single nearby bird is often easy to recall years later. Roger Tory Peterson tells of the day when he was eleven years old and touched a bundle of feathers clinging to the trunk of an oak, only to discover with a shock—as the startled bird whipped its head around and stared at him—that it was alive. The mass of feathers was a sleeping Northern Flicker, but the shock—the connection—was so great that Peterson saw it as an event that directed the course of his life. Bird artists and bird scientists all have memories of their first great connection with birds, and of many later ones as well. They bring some of those connections to us, sometimes as art, sometimes as research, sometimes as both. Science Art, at its best, conveys these connections as though they are happening again.

APPENDIX 1 Timeline Linking Art, Technology, and the Study of Birds

Images document the slow rise of ornithology as a bona fide science. They record the transition of birds from objects of interest, sometimes venerated, in prehistory and the ancient past, into subjects of organized study in contemporary times. They register the discoveries made on the European mapping expeditions in the 1500s. They confirm the burgeoning European specimen collections of the mid-1500s and the growing roster of species named by Europeans in the 1600s. They convey the developing protocols that allowed researchers to organize birds (as well as other animals and plants) into biologically meaningful categories, which led in the 1700s to wide acceptance of the binomial nomenclature that gives each species a unique name. And they chronicle how the availability of unique names led to a rapid growth in the Western market for bird books, especially illustrated accounts of exotic (nonnative) species.

In the 1800s, illustrated books filled with paintings of birds, now named and grouped by similarities, helped readers see how evolutionary theory lent order to the background noise of natural philosophy. Developments in the understanding of birds that were reported in the scientific literature and captured on canvas eventually brought to light the vulnerabilities of bird populations; awareness of those vulnerabilities led, by the late 1800s, to the development of a conservation agenda. Then, with the publication of field guides filled with small, easily recognized paintings of each species, and the production of affordable binoculars, the barriers holding back public involvement in bird conservation efforts slipped away. Today the constituency of activist birders and "citizen scientists" is very large; their concerns are increasingly documented by images, and the results of their efforts are more impressive than early conservationists might have thought

possible.[1] On the other hand, an equivalent expansion has taken place in the number and severity of insults to the avian environment that today's conservationists must confront and work to diminish.

The timeline below provides a context for the art discussed in this book.[2] It lists major figures who produced bird art or made advances in bird science, as well as major shifts in technology that affected one or both arenas. The entries in black type refer to artists, trends, or events in art. Artists are typically listed at the midpoint of their career. Each artist with an image featured in this book is listed in red at the date of the cited work. The section covering the past half-century contains only one artist, however, because only deceased artists are included here. When the artist is unknown, the title of the work is listed in red.

The entries printed in blue refer to innovators, trends, or events in science and technology. Innovators are listed either at the midpoint of their career or at the point of their major contribution.

The remaining entries, printed in green, refer to cultural events that relate to birds, the production of bird art, or both—for example, the expeditions that returned with exotic specimens during the Age of Exploration. Individuals who made contributions to both art and science are coded according to their major influence.[3] Please note that given the availability of scholarly resources to us, this timeline emphasizes Western scientific, cultural, and artistic events, despite efforts by the authors to include important advances from non-Western sources.

Era/Date/Midpoint	Events and Individuals Influencing the Production of Bird Art
Paleolithic Age	
c. 30,000 BCE	*Owl in Chauvet Cave,* France (Plate 2).
c. 30,000–17,000 BCE	*Owls in Les Trois Frères Cave,* France (Plate 24).
c. 17,000–16,000 BCE	*Auk in Cosquer Cave,* France (Plate 17).
c. 15,000–10,000 BCE	*The Shaft in Lascaux Cave* (detail), France (Plate 31).
Neolithic Age	
c. 8000? BCE	*Netted Ostriches on a Libyan Outcrop,* Libya (Plate 10).
Ancient Civilizations	
c. 3500 BCE	Pictographic writing appears in Sumeria.
	Phoenicians discover glass while cooking on sand.
c. 3000 BCE	Hieroglyphic writing appears in Egypt.
c. 2500 BCE	*Statue of King Chefren,* Egypt (Plates 4 and 5).
c. 2000 BCE	Papyrus "books" appear in Egypt (e.g., *Book of the Dead*).
c. 1500–1100 BCE	Chinese images include stylized decorative birds, among them, bronze vessels in the shape of owls.
early 1400s BCE	*Pelicans from a Wall Painting in the Tomb of Horemheb,* Egypt (Plate 12).
1100–250 BCE	Chinese images include Mandarin Ducks on bronze and pottery vessels.
c. 700–650 BCE	The use of coins is adopted in Lydia and spreads to Greece.
500 BCE	Herodotus, a Greek historian, includes natural history in his writings (see Plate 33).
c. 412–411 BCE	*Ten Drachma Silver Coin,* Sicily (Plate 25).
c. 400 BCE	What Westerners come to call arabic numerals appear in India, although the date is uncertain (they are later conveyed to Europe by Arab scholars), with the zero appearing around the ninth century.
360–343 BCE	*The God Horus Protecting King Nectanebo II,* Egypt (Plate 6).
330 BCE	Aristotle, a Greek who is considered by some the father of biological thought, derives a crude sense of evolution and purportedly writes 400 books, of which 48 remain. Of his multivolume *Historia animalium,* four volumes are on birds. Aristotle recognizes 140 bird species, and his list is a source for other early listers of birds, including Pliny, Frederick II, Leonardo da Vinci, Pierre Belon, and Ulisse Aldrovandi.
305–30 BCE	*Coffin for an Ibis,* Egypt (Plate 33).
250 BCE	Asoka introduces wildlife protection in India.
before 210 BCE	The Chinese general Meng Tian invents by this time a writing brush for use on silk.
before 200 BCE	Mayan hieroglyphic writing is used in Guatemala.
200 BCE	King Devanampiyatissa establishes a wildlife sanctuary in Sri Lanka.
	Paper is invented in China, and its use becomes widespread.
c. 95–55 BCE	Lucretius proposes a "ladder of Nature" based partially on Aristotle's thought and might be seen to have anticipated modern evolutionary theory.
50 CE	Pliny the Elder, a Roman compiler, produces a 37-volume *Historia naturalis,* one volume of which is on birds. Repeatedly copied, it is among the first works to be printed, in 1469. Pierre Belon uses it in the 1500s, but it is increasingly debunked (see Plate 33).

170	Galen, an anatomist living mostly in Rome, is the last major biologist of antiquity. He writes 256 books, 131 of which are on medicine. Galen rejects Aristotle's ideas on evolution and is highly respected during the Middle Ages.
c. 225	Aelian, a Roman compiler, produces *On the Characteristics of Animals* in Greek. It is a collection of animal stories based primarily on the work of others, with much of the bird material taken uncritically from work by Alexander of Myndos and Plutarch.
c. 275	C. Julius Solinus, a Roman compiler, produces *Collectanea rerum mirabilium*, taken mostly from Pliny.
c. 300	Wang Yi of China writes a treatise entitled "A Guide to the Study of Painting: Birds and Insects."
370	*Physiologus* is compiled in Alexandria or Syria.

500

511	Xie He, a Chinese painter, is credited with founding a scientific theory of Chinese painting, although his writing is based on earlier rules.
early 500s	The first illustrated manuscripts of the Bible appear.

600–700

early 600s	Isidore of Seville, the bishop of Seville and an encyclopedist, compiles *Etymologia*, which includes a chapter on birds.
600s	Paper is introduced into the Middle East from China.
c. 600–650	*Purse Lid from the Sutton Hoo Ship Burial,* England (Plate 13).
c. 600–900	In China, ink on carved wooden blocks allows for multiple transfers of an image to paper.
676	Cuthbert of Lindisfarne protects birds on the Farne Islands in the United Kingdom.

800

813–833	The Caliph Ma'mum constructs Bayt al-Hikma, the "House of Wisdom," a university-like campus where works of Greek scholars and others are translated into Arabic. It is destroyed along with other libraries during the Mongol capture of Baghdad, purportedly turning the Tigris black with ink from the vast number of books flung into the water.
868	Printed books appear in China.
late 800s	*St. Gregory with Scribes,* Carolingian (Plate 7).

900

c. 900	Chu Hui creates bird and flower paintings in China. He is considered the father of that genre.
934	Huang Quan, a Chinese painter, paints without the use of outlines. He produces *Animals Sketched from Nature (Xiesheng Zhenqin Tu)* (Plate 19).
960–1280	Painted birds produced in China are now seen beyond its borders.

1000

c. 1040	Movable type made from porcelain appears in China.
1041	Movable type made from clay appears in Korea.

1100

1100–1126	Hui Zong, Chinese emperor and artist, reigns. He initiates academic art.

1100s	Arabic numerals reach Europe.
	Paper is introduced into Spain. It is produced in quantity in Germany and Italy.
1200	
c. 1201	Li Anzhong, a Chinese artist, portrays active birds.
c. 1230	Movable metal type appears in Korea.
c. 1230–1240	*Barnacle Goose,* England (Plate 27).
1241	Books printed using movable type appear in Korea.
c. 1248	Frederick II, Holy Roman Emperor, writer, and natural historian, is the first eminent European ornithologist. In his spare time he builds animal parks and practices falconry, which was brought to Sicily by the Saracens and spread from there. Frederick II has Michael Scott translate Aristotle's long-lost work on animals from Arabic into Latin (1230) and writes *The Art of Falconry* (*De arte venandi cum avibus*), which includes 900 marginal images showing raptor biology. It is the first systematic bird book based on observation where life histories are biologically sound.
c. 1260–1270	Albertus Magnus, a Dominican friar and teacher, writes the 26-volume *De animalibus,* one volume of which is on birds. His bird survey includes comments on Aristotle's survey of 140 birds along with information on German birds.
1286	The first wearable eyeglasses are made in Italy.
1300	
1300s	*Page from a Monk's Drawing Book,* England (Plate 20).
1309	Europeans make paper. A paper mill appears in France in 1338 and in Germany in 1390.
c. 1334–1348	Andrea Pisano **produces** *The Art of Flight, Daedalus* (*Il Volo di Dedalo*) for the Bell Tower in Florence, Italy (Plate 21).
1400	
c. 1410	Oil paint based on linseed oil and used with a stable varnish appears in Europe.
1414	Noami, a Japanese painter who incorporates the spirit of Chinese art, paints murals of flying birds.
1418	Woodcuts are used in Europe both for religious images and for tarot cards.
c. 1420–1460	Prince Henry the Navigator of Portugal encourages exploration.
1423	Woodcuts are used in books in Europe (xylography); Albrecht Pfister produces a Bible.
c. 1440	Johannes Gutenberg of Germany invents a printing press with movable metal type.
1455	Sesshu, a revered Japanese painter, brings Ming and Yuan (Chinese) influences to Japan.
	Block books, printed from hand-cut wooden blocks (one block per page), which were developed in China hundreds of years earlier, appear in Europe.
1457	Color printing appears in Europe with the publication of the *Mainz Psalter.*
1461	The first printed book with woodcut illustrations appears in Europe: *Der Edelstein,* by Ulrich Boner, printed by Albrecht Pfister of Bamberg, Germany.
1465	In Germany, drypoint engraving (made by scratching lines into a plate with a pointed tool) is invented.

1486	Bartholomew Diaz, sailing for Portugal, rounds the Cape of Good Hope.
1492	Christopher Columbus, sailing for Spain, lands in the Bahamas.
1493	Daniel Hopfer, in Germany, produces what are widely believed to be the first etchings used in printmaking.
1497–1499	Vasco da Gama sails from Portugal to India. The trade that opens up with China afterward includes lacquer scroll paintings and porcelain. Live birds are also carried to Europe, where artists paint them.
1500	
1500	Albrecht Dürer, the premier German artist of his time, knows Leonardo da Vinci and brings Renaissance ideas and techniques to northern Europe. Dürer's study of the wing of a roller is often copied or imitated (see Plates 62 and 63).
c. 1505	Leonardo da Vinci, an Italian painter, observer of nature, and student of mechanics, optics, and animals, studies flight relentlessly. He designs *Model of Flying Machine Driven by the Force of Man* (Plate 43) around 1505. One of his few paintings, *Leda and the Swan*, includes an iconic bird. The painting is lost, but the preparatory drawing remains, and a copy of the painting by Cesare da Sestro is thought to be faithful.
1513	Urs Graf, in Switzerland, produces two etchings, one of which dates from 1513, and invents a woodcut technique whereby white lines create an image on a black background.
	Vasco de Balboa, sailing for Spain, reaches the Pacific Ocean.
1516–1530	Tomé Pires, a Portuguese trader, leads a mission to China. The emperor refuses to grant Pires an audience when he arrives in Nanjing, and drives off the Portuguese. Pires dies in prison in 1524, and the prohibition of foreign traders is not lifted until 1530.
1520–1522	The Portuguese Ferdinand Magellan sails around the globe.
1533	Hans Holbein the Younger, a mainstream German portraitist who works for years in Basel, Switzerland, and London, becoming court painter to Henry VIII, includes falcons in two portraits and a starling in a third. He paints other birds as well, successfully conveying their essence (see the frontispiece).
1543	Portuguese traders reach Japan by accident, blown off course during a typhoon. They repair their ship and return to their Portuguese base in Malacca.
1544	William Turner, an English writer, publishes the work of the recently deceased Gilbertus Longolius in *Dialogus de avibus et earum nominibus graecis, latinis, et germanicis* (1544), which deals with the names the "ancients" gave to birds. He also publishes *Avium praecipuarum, quarum apud Plinium et Aristotelem mentio est, brevis et succincta historia* (A Short and Succinct Account of the Principal Birds Mentioned by Pliny and Aristotle; 1544), which is applauded by Merrem 244 years later and is valued even today as a work that describes birds as they are, without embellishment or speculation.
1555	Pierre Belon, a well-traveled French naturalist, studies comparative anatomy by dissecting many birds, which he lists by affinity based on Aristotle's system. His book on the natural history of birds, *L'histoire de la nature des oiseaux* (1555), with its hand-colored woodcuts, is overshadowed by the simultaneous publication of Conrad Gesner's *Historia animalium,* but his studies of human and bird skeletal homologies continue to illustrate zoology books. Belon adds an extra pair of bones to the leg portion of his bird skeleton, but his drawing still marks an improvement over previous studies in anatomy.

1555	Conrad Gesner, a Swiss-French physician, writer, and natural historian, publishes a four-volume descriptive encyclopedia, *Historia animalium*, one volume of which is on birds. Although he lists birds according to Aristotle's system, he sets two new standards: by assembling data from a network of scholars, he energizes the effort to broaden the coverage of birds beyond local regions, and he provides one woodcut for each species of bird included, for a total of 217 images. New editions of his work are published over the next 200 years, serving as a source for a long line of authors.
1568	Pieter Brueghel the Elder, a Flemish painter, depicts active people and aspects of human behavior, sometimes including birds as usually small but important elements in the narratives. *The Peasant and the Nest Robber* (Plate 34) conveys the idea of conservation.
1575	Jacopo Ligozzi, an Italian artist who portrays rarities, is appointed court painter for Grand Duke Ferdinand II. He also works for the Medici family. He illustrates Aldrovandi's *Ornithologia* in 1599–1603.
1577–1593	John White, an English artist who sails with the founding colonists of Virginia, returns home with Sir Francis Drake, then sails back to the colonies with Sir Walter Raleigh, who leaves him there as governor of Roanoke. White paints 25 recognizable New World bird species among other fauna, and also paints native people.
1590	Two Dutch spectacle makers, Zaccharias Janssen and his father, Hans, invent the microscope.
1599–1603	Ulisse Aldrovandi, Italian polymath, professor at the University of Bologna, and collector, publishes his three-volume *Ornithologia* (*Ornithologiae hoc est de avibus historiae libri XII*). Although it lacks original material, it is the best ornithological work of the day and remains a standard for more than a century. Unlike Gesner, who lists birds alphabetically by common name, Aldrovandi creates a system; even so, the best feature of the book may be the artwork— *Ornithologia* contains 685 woodcuts from original colored drawings by Jacopo Ligozzi—and the worst may be his inclusion of the lore of the Barnacle Goose (Plate 27).
1600	
1607	Like Georg and Jakob Hoefnagel and other animal artists, Roelandt Savery, a Flemish painter of animal portraits, works in Prague for Rudolf II, whose zoological gardens are a source of subjects. Savery's Dodo paintings are well known.
1608	Hans Lippershey, born in Germany but relocated to the Netherlands, constructs a telescope.
1609	The Italian astronomer Galileo Galilei improves the telescope, increasing its magnification, and proposes a scientific approach to the analysis of the natural world.
	Peter Paul Rubens, a baroque painter in northern Europe, establishes an extensive studio. He includes birds in numerous paintings; the birds are often added to the paintings by a collaborator, Frans Snyders.
c. 1609	The Persian miniaturist, Isfahan Allah Habib, illustrates *The Language of Birds,* a book-length poem about Sufism by Fard-al-din (c. 1609). In the poem a hoopoe answers questions posed by 30 birds.

1618	Frans Snyders, a Flemish painter, specializes in realistic, active birds and bird trophies during the Golden Age of Dutch and Flemish painting, when still lifes, larder art (images portraying domestic scenes that include hunted, dead birds among other bounty), and trophy pictures were in demand. Snyders often collaborates with other artists: he puts birds in the paintings of Peter Paul Rubens and Jacob Jordaens, and they put people in his.
1638–1639	Rembrandt van Rijn, a prolific Dutch painter, features birds (both dead and alive)—including a bittern, a peacock, an eagle, and a falcon—in several works, and he draws birds of paradise (see Room 7).
1644	The Royal Academy of Sciences is founded in Paris.
1648	Georg Marcgraf, a German physician and naturalist, collects natural history specimens in Brazil from 1637 to 1643 but dies a year later, at age 34 (probably of malaria), prior to refining his notes into publishable material. Johannes de Laet, a scholar and the director of the West Indies Company, decodes Marcgraf's notes and publishes a folio, *The Natural History of Brazil*, in 1648. After de Laet's death, Willem Piso takes over, and he publishes *De Indiae utriusque re naturali et medica* ten years later. Piso does not mention Marcgraf in the book and deletes much of his material, but the volumes based on Marcgraf's work, which introduce 133 species new to Europeans, remain the definitive coverage of Brazilian fauna and flora for 150 years.
1650–1653, 1657	John Jonston, who was born in Poland of Scottish descent and received his doctorate at Leiden, the Netherlands, publishes *Historiae naturalis de avibus libri* in six parts. The book on birds is a popular work with good illustrations by Matthaeus Merian the Younger (mostly after woodcuts found in Gesner's and Aldrovandi's books) but contains very little original material. An English translation appears in 1657.
1654	Carel Fabritius, a Dutch mainstream painter who was a student of Rembrandt's, leaves perhaps eight paintings to posterity, including *The Goldfinch*, a trompe-l'oeil painting of a pet chained to a perch on a wall.
1655	Ole Worm, a Danish natural philosopher and professor of medicine, like Aldrovandi, assembles a natural history collection as a teaching aid. His collection in Copenhagen is known throughout Europe, and a four-part catalogue, *Museum Wormianum*, which includes illustrations from a number of publications (Gesner's and Aldrovandi's, among others), is published in 1655, a year after his death. The catalogue contains 22 pages on birds and a frontispiece of his "collared" pet Great Auk (see Room 8).
	Charles Lebrun, a French historical painter, founds the Académie Royale and serves as the first director of the Gobelins Manufactory, which makes tapestries. His well-known portrayal of Daedalus attaching wings to Icarus is sometimes considered to represent the theme of Abraham's sacrifice of Isaac.
1662	The Royal Society is founded in London.
1665	The first periodical in Europe, *Journal des sçavans*, is published, followed several months later by *Philosophical Transactions of the Royal Society* in England.
1666	Christopher Merrett, a British physician, publishes *Pinax rerum naturalium Britannicarum*, considered by some the first classical British bird book. In it the classification of species follows the systems of Aldrovandi and Jonston.
1675	David Klöcker Ehrenstrahl, a German artist who paints portraits and history paintings, becomes a court painter in Sweden. His biologically instructive painting of lekking grouse, *Blackcocks in Springtime (Orrspel)* (Plate 68), is highly praised.

1676	Francis Willughby, an English zoologist, teams up with John Ray, an English botanist, to describe the entirety of nature. Their classification system, based on habitat, size, and appearance, supersedes the classification system established by the encyclopedist Aldrovandi and paves the way for Mathurin-Jacques Brisson; Georges-Louis Leclerc, comte de Buffon; Carolus Linnaeus; and scientific classification generally. Willughby, who dies in 1672, leaves a life income to Ray, who publishes their work as *Ornithologiae* in 1676. The book describes 230 British birds.
1700	
1702	François Desportes, a popular mainstream French artist, specializes in still lifes, larder art, and images of hunting trophies. He excels at portraying plumage. Louis XIV names Desportes the official painter of the Royal Hunt and commissions him to record his menagerie. Desportes is often compared with Jean-Baptiste Oudry and is also associated with the Gobelins and Beauvais tapestry factories, which produce Europeanized versions of flora and fauna of the New World.
c. 1705	Maria Sibylla Merian, a German natural history artist specializing in insects, travels to Surinam, publishing on her return *Metamorphosis insectorum Surinamensium* (c. 1705). It includes some birds, showing aspects of their biology. Merian mixes accuracy and aesthetics, setting a standard for scientific illustration that influences, for example, Eleazer Albin, Carolus Linnaeus, Mark Catesby, and Charles Willson Peale.
1709	Kaibara Ekiken, perhaps the first major Japanese ornithologist, publishes the 21-volume *Yamato honzo* (The Natural History of Japan) in 1709. The volume dedicated to birds covers 99 species systematically and sets the standard in Japan for a century.
1711	Carl Wilhelm de Hamilton, a Scottish-Flemish animal painter, produces two versions of *The Parliament of Birds* based on a poem by Chaucer that describes avian mate-selection as an orderly event occurring on Saint Valentine's Day. The painting reproduced here (Plate 29) shows the contemporary familiarity with local diversity.
1724–1755	Jean-Baptiste Oudry, a French artist who was initially a portraitist, becomes a highly productive painter of hunting scenes and other pictures of birds and mammals after Louis XV has him paint the animals in his menagerie. Like Desportes, he is involved with tapestry.
1731–1738	Eleazar Albin, a German painter and writer known for paintings of insects and objects of curiosity, produces books through subscription, including a bird book with hand-colored etchings. Albin's bird-on-a-branch format with occasional depictions of backgrounds and food sources becomes widely used. His three-volume *A Natural History of Birds* (1731, 1734, 1738) includes 10 North American species. It is illustrated in color with 306 copper engravings, but owes perhaps too much to Francis Willughby and John Ray. He also writes *A Natural History of English Songbirds* (1737), the first book to include many eggs. Some copies are hand-colored.
1731, 1743	Mark Catesby, an English writer and illustrator financed by Sir Hans Sloane for travel to North America, publishes *The Natural History of Carolina, Florida and the Bahamas,* with about 100 illustrations of birds. This publication set the standard for American natural history books until Alexander Wilson and John James Audubon produce theirs. Catesby etches his own illustrations, and he uses a large page, allowing him to produce life-size representations of birds in their correct habitat with ecologically correct prey, vegetation, etc.

1734–1765	Albert Seba, a well-known Amsterdam collector, privately publishes an unorganized catalogue of his collection, *Accurate Descriptions of the Most Richly Endowed Treasury of Nature and an Illustration with the Most Skillful Pictures, for a Universal History of the Physical World*. The illustrations are better than the text, and some unidentified species could represent extinctions.
mid-1700s	François Boucher, who paints *The Discreet Messenger* (Plate 15), is a mainstream French rococo painter. He specializes in decorative genre scenes that range from depictions of animals to religious narratives.
	Pyon Sang-Byok, a Korean painter, is a member of the official Painting Bureau. He is valued for his realist works, especially those in which he departs from the official academic style.
1743–1764	George Edwards, an English natural historian, writer, and illustrator, paints specimens for Sir Hans Sloane, president of the Royal Society, among others, and eventually writes multivolume bird books illustrating mostly captive exotics, many from North America. Linnaeus names 350 birds based on Edwards's descriptions. Edwards begins publishing his *Natural History of Uncommon Birds* in 1743 and completes the series in 1764; the volumes from 1758 are entitled *Gleanings of Natural History*. His work is a popular success, and the birds are accurate though not systematically arranged. Edwards also prepares a catalogue of Sir Hans Sloane's collection.
1751–1772	The Frenchman Denis Diderot writes his encyclopedia (see the Mezzanine).
1756	Sir Hans Sloane, a British naturalist and collector, helps finance the work of Mark Catesby. By 1756 his avian collection surpasses all others in Europe save René-Antoine Ferchault de Réaumur's. On January 15, 1759, his library and avian collection of 1,172 items, including bird parts and eggs, opens to the public as the British Museum.
1758	Carolus Linnaeus, a Swedish physician, professor, naturalist, and botanist, publishes the 10th edition of *Systema naturae* in 1758. Organized according to two-part Latin names, it marks the official beginning of binomial nomenclature for birds. Linnaeus's knowledge of birds is limited, and he does not access the superior collections of Europe. His bird coverage partially follows Francis Willughby and John Ray, who classified birds using bill, foot, and body size, a system more natural than the one Linnaeus eventually devises. Linnaeus's system of bird classification is also inferior to Brisson's, published in 1760, but Brisson's benefits future editions of Linnaeus's book.
1760	René-Antoine Ferchault de Réaumur, a French naturalist, builds the largest avian collection in Europe (he dries birds in ovens still warm after bread baking) and hires Brisson to curate it. Brisson publishes the collection catalogue in 1760. Réaumur bequeaths the collection to the Académie des Sciences, an instruction that the king (with the support of Buffon) ignores, and it is transferred to the royal collection.
	Mathurin-Jacques Brisson, a French ornithologist, publishes a six-volume catalogue, *Ornithologie*, that covers all 1,500 bird species known to Western naturalists, plus 320 new forms found in collections. François-Nicholas Martinet illustrates the series, which is based primarily on Réaumur's collection, the most complete European one to date. Brisson, curator of the Réaumur collection, devises a classification system that divides birds into 115 genera (a division more accurate than Linnaeus's) and groups them into 26 orders (20 more than Linnaeus has), building a system more like the one in international use today than any other devised during the next 80 years.

1761–1766	Thomas Pennant, a leading British ornithological writer, publishes the four-volume *British Zoology* (1761–1766), reinvigorating interest in native birds and boosting ornithological writing in Britain. He also publishes *Indian Zoology* (1769) and the bird volume of *Arctic Zoology* (1784–1785). Both Pennant and the other leading British ornithological writer, John Latham, fail to follow the Linnaean system of binomial nomenclature and use only English bird names.
1764	Morten Thrane Brunnich, considered by many to be the founder of Danish zoology, names new birds according to the Linnaean system of binomial nomenclature, making him, along with Peter Simon Pallas and Eric Pontoppidan, among the first zoologists to understand the value of the method. In 1764, Brunnich publishes *Ornithologia borealis*, separating two murres into their own species, one of which bears his name.
1764?	*Untitled* [The banding of a heron] shows the early use of metal bird bands (Plate 35).
1765–1773, 1770–1783	Georges-Louis Leclerc, comte de Buffon, publishes the nine-volume *Histoire naturelle des oiseaux*, which includes 1,008 hand-colored engravings depicting 1,239 birds. The series, which becomes an extremely popular work, elevates the royal collection in Paris under his care to a major research center for the natural sciences. Between 1765 and 1773, Buffon produces *Planches enluminees*, which contains 1,939 hand-colored bird engravings. He is remembered for publishing the first illustration of a species that had gone extinct—the Dodo—in his 44-volume encyclopedia, *Histoire naturelle, générale et particulière* (1749–1804).
1767, 1777	Johann Reinhold Forster, a German naturalist, publishes work on the flora and fauna of southern Russia in 1767, utilizing the Linnaean system and identifying 74 bird species. Forster goes on to study and compile information on birds from India, China, and the American colonies. Together with his son George, he replaces Joseph Banks on Captain James Cook's second voyage of exploration (1772). Although Forster's *A Voyage Round the World* (1777) is not a success, the Forsters' joint contribution to science overshadows that of Banks. And although financial ruin leads to a short stay in debtor's prison and the dispersal of the Forster collection, Forster's monographs are still considered superlative.
c. 1768	Joseph Wright of Derby, a mainstream British artist, is linked to but not a member of the Lunar Society, a club whose members are interested in scientific discoveries. Wright depicts a Lunar Society event in a well-known painting of a Sulphur-crested Cockatoo (*An Experiment on a Bird in the Air Pump*, Plate 44).
1768–1771	Sidney Parkinson, a Scottish artist, accompanies Joseph Banks aboard the *Endeavor* on Captain James Cook's first global voyage (1768–1771), producing at least 1,300 images.
1769	Joseph Banks, a botanist, joins Cook on his first global voyage (1768–1771) at his own expense, turning his attention to fauna while at sea and observing the Transit of Venus on Tahiti. In 1772 he plans to take Daniel Solander (a pupil of Linnaeus and a British Museum staff member) around the world on Cook's second voyage, but he is denied passage.
1770–1776	Maruyama Okyo, a Japanese artist who advocates working from nature, produces bird images with greater realism than is found in paintings by most Japanese contemporaries.
1777	Colson, a French artist who writes works on perspective and is elected to the Dijon Academy of Sciences and Academy of Arts and Letters, in 1777 paints a *Portrait of Balthazar Sage*, a pharmacologist who has subjected two of three passerines to a toxic agent (Plate 22).

1780–1830	Copper engraving gives way to stipple engraving. The 145 color prints by Barraband in Levaillant's *Histoire naturelle des perroquets,* published in 1801–1805, are examples.
1781	Matsumura Goshun of Japan advances the naturalism of Maruyama Okyo, forming the Shijo school.
1781–1785	John Latham, a leading British ornithologist, is also a physician, museum owner, and cofounder of the Linnaean Society. He publishes multivolume bird books illustrated with simple frozen forms showing plumage. *A General Synopsis of Birds* (1781–1785), including 106 plates drawn, etched, and colored by Latham, was followed by two supplements. *Synopsis* was initially valued for including the 200 species obtained on Cook's second and third voyages but was criticized for the failure of the collectors to provide locality labels for specimens. Latham is slow to begin latinizing names; he waits until publication of the *Index ornithologicus* (1790), which includes all 3,000 birds known in the West at the time (compared with the 930 that Linnaeus listed in 1758) and names hundreds of new birds, especially from Australia.
1782	George Stubbs, who painted *Farmer's Wife and the Raven* (Plate 8), is a virtually self-taught mainstream British painter and engraver who specializes in animal images, deriving accuracy from detailed study and dissections.
1784	Charles Willson Peale, a portrait painter and naturalist, founds the Peale Museum (now part of the Philadelphia Museum, and later the Academy of Natural Sciences in Philadelphia), which expands in the next century by adding specimens from expeditions, including Major Stephen H. Long's 1819–1820 expedition to the Arkansas River, on which Peale's son Titian serves as preparator (he is in charge of preparing specimens for the collection) and artist and Thomas Say as zoologist.
1788	Frederick II's book *The Art of Falconry* (c. 1248) is rediscovered by both Johann Gottlob Schneider and Blasius Merrem; the former produces a highly valued annotated edition.
1789	Jacques Barraband, a French bird artist, not only introduces the color-printed copperplate technique but also circulates prints by Hokusai around Paris that eventually influence French impressionists. Barraband produces images for tapestry and porcelain makers and illustrates books for other bird writers, including Levaillant.
	Gilbert White publishes *The Natural History of Selbourne,* including notes on 120 or so local bird species for this region 50 miles (80 kilometers) south of London. The book consists primarily of his letters to Thomas Pennant and Daines Barrington and influences such later nature writers as John James Audubon, Margaret Morse Nice, Rachel Carson, and Roger Tory Peterson. Like George Edwards and Thomas Pennant, he records migration dates of possible use in contemporary studies of global climate change.
1790–1808	François Levaillant, a native of Surinam, goes to South Africa in 1781, sent by Jacob Temminck, and returns in 1784 with more than 2,000 skins of birds. Various books follow—the popular *Voyage de Monsieur de Le Vaillant dans l'intérieur de l'Afrique in 1790;* a sequel, *The Second Voyage* (1795); and *Histoire naturelle des oiseaux d'Afrique* (1796–1808), illustrated with copperplate engravings by J. L. Reinold. Levaillant's eloquent accounts of African birds include descriptions of some 50 new species. One of Buffon's star students, he defends Buffon's opposition to Linneaus's systematics.

1791	William Bartram, though primarily a botanist, is also known as America's first scientific ornithologist and first ecologist. He helps elevate nature writing to the status of literature; his *Travels through North and South Carolina* (1791) includes a *Catalogue of the Birds of North America* that provides information on the habits and migration of 215 species. He corresponds with Pennant and Edwards, sending them specimens, and is a great asset to his new Philadelphia neighbor Alexander Wilson, whom he meets in 1803, opening his library to him and fostering his interest in ornithology.
1795	John Abbott, an English naturalist and artist, emigrates to the American colonies, settling in Georgia to collect, describe, and illustrate the flora and fauna of the Savannah River valley. His five copies of *The Birds of Georgia* are never published, but his specimens and 1,400 illustrations are sources for others, including John Latham.
1796	The Bavarian author Aloys Senefelder invents lithography.
1797	Thomas Bewick, a British artist, writer, and naturalist, publishes a two-volume *History of British Birds*, known for its 448 wood engravings and their depicted backgrounds, which leave a record of preindustrial British landscapes. The engravings, less expensive than hand-painted plates, make Bewick's book widely accessible and help popularize natural history. He is respected by Audubon (who names a wren for him) and praised by William Wordsworth, Charlotte Brontë, John Ruskin, and others.
1799	Louis Robert of France invents the Fourdrinier machine for sheet paper making.
1800	
1803	In North America, the Lewis and Clark expedition reaches the Pacific.
	Ono Ranzan publishes the 48-volume *Honzokomoku keimo*, which has two volumes on birds. The book improves on Kaibara's work a century earlier, as well as other natural history works on Japan published during the intervening years.
1805–1809	Louis Jean Pierre Vieillot emigrates from France to Santo Domingo in 1780 but, on the rise of civil unrest there, relocates around 1792 in the United States. Six years later, destitute, he returns to Paris. He publishes *Histoire naturelle des plus beaux oiseaux chanteurs de la Zone Torride* (1805–1809) and between 1812 and 1820 studies the expanding exotic bird collection in the Paris Museum. Vieillot, though not considered a scientist, introduces 26 new generic names. He is overshadowed by Buffon and Georges Frédéric Cuvier and rejected by contemporary ornithologists but gains favor after his death.
1808–1811	Pauline de Courcelles (Madame Knip), a student of Barraband's, publishes *Histoire naturelle des Tangaras, des Manakins et des Todies* in 1805; it includes 72 of her illustrations of exotics. From 1808–1811 she collaborates with Coenraad Jacob Temminck on a pigeon book. (Temminck provides the text, but she takes full credit in the by-line. He later collaborates on pigeons again, from 1838–1843, working this time with Florant Prévost, assistant naturalist at the Paris Museum.)
1808–1833	Alexander Wilson, a Scottish-American weaver and schoolteacher, makes good use of William Bartram's library in Philadelphia and, in 1808—on his way to becoming the "father of American ornithology"—publishes the first volume of *American Ornithology*, which he also illustrates. The last two volumes of the nine-volume work are completed after his death, by George Ord in 1833. Wilson's art is later overshadowed by Audubon's, but not his observations or writing.

1809	Jean Baptiste de Lamarck, an assistant to Buffon, is credited with coining the term "biology." Lamarck publishes *Philosophie zoologique* in 1809, asserting that all systematic categories above the species level are artificial (a debate from the 1300s). Only after his death does his view that use determines form (e.g., the neck of the giraffe and the webbed foot of the duck are both the result of stretching) make his name widely known.
1810	Alexandre Isidore Leroy de Barde, a French-English natural history painter, relocates during the French Revolution and produces six large trompe-l'oeil paintings of objects from natural history collections, including *A Collection of [39] Foreign Birds* (1810) and another work featuring 23 avian species. Both are exhibited in London in 1814, and all six are exhibited in the Paris Salon in 1817.
	Frederick Koenig, who moved from Germany to England a decade earlier, improves the printing press.
1811	Peter Simon Pallas is a German who lives in Russia for 44 years and whose major contribution to biological science, *Zoographic Rosso-Asiatic,* is published the year of his death. He blends the rigid systematics of Linneaus and the mutability of species of Buffon.
	Carl Illiger, a German ornithologist, is curator of the University of Berlin's Natural History Collection, which is rich in specimens from Brazil, North America, and Australia. He publishes *Prodomus systematis mammalium et avium,* in which he revises the Linnaean system. He divides birds into 7 orders, 41 families, and 147 genera, a division that fosters use of the concept of family and leads to studies of relationships, and he supports the idea that climate and other unknown influences promote variation.
1812	The Academy of Natural Sciences opens in Philadelphia complete with a journal. It is associated with the Philadelphia Museum and is the first U.S. museum to contain an ornithology collection.
1813	Blasius Merrem, a German attorney, is also an ornithologist with a strong interest in anatomy. He discovers the air-sac system in birds by injecting wax through the trachea and publishes his results in *On the Air-Apparatus in Birds* in 1783. His main contribution to ornithology, although it is said to have had little impact, is the publication of fragments of *An Outline of the General History and Natural Classification of Birds* (1813), in which he produces the first modern classification of birds.
1817	Georges Frédéric Cuvier is an early follower of Buffon, whose catastrophism asserts that Earth undergoes periodic life-destroying local catastrophes, whereupon species are replaced by immigrants. Cuvier considers species invariable and repudiates Lamarck's hypothesis of evolution. His *Le régne animal* (1817) includes a classification of birds.
1820	Coenraad Jacob Temminck takes over the 1,100-specimen collection accrued by his father, Jacob Temminck, treasurer of the Dutch East India Company. His *Manuel d'ornithologie, ou Tableau systématique des oiseaux qui se trouvent en Europe* (1815), remains the standard work on European birds for a long time. In 1820 he becomes the first director of the Rijksmuseum in Leiden, where he remains until his death in 1858.
1824 or 1826	Joseph Nicéphore Niépce develops the first photographic process, heliography, but it requires eight hours or more to take a picture.
1825	J. P. Lemiere invents the first binocular telescope, which consists of two similar telescopes, one for each eye, mounted on a single frame.

c. 1825, 1829?	John James Audubon, a French-American bird and mammal painter and the artistic father of American ornithology, explores the eastern United States. Audubon creates life-size images of U.S. species in his nine-volume elephant-folio (c. 30 × 40 inches; 76.2 × 101.6 centimeters) *Birds of America* (1840–1844), which includes 435 hand-colored engraved plates based on watercolors illustrating 487 species positioned in lifelike poses (e.g., *Northern Mockingbird [Mocking Bird]*, Plate 62, and *Brown Thrasher [Ferruginous Thrush]*, Plate 63).
1830–1860	Hand-colored lithography is widely used for bird and other illustrations, allowing artists to work directly on stone and eliminating the role of the engraver.
1830–1885	The study of birds using the Western scientific method develops in India. Major General Thomas Hardwicke publishes drawings of birds, chiefly selected from his own collection, in *Illustrations of Indian Zoology,* which is edited by John Edward Gray of the British Museum beginning in 1830. Edward Blyth, who serves as curator of the Asiatic Society's museum in Calcutta for two decades, mentors Allan Octavian Hume, who builds a large collection of Indian bird skins and eggs (63,000 skins, 19,000 eggs) and donates it to the British Museum in London in 1885.
1831	George Robert Gray, an ornithologist, becomes curator of the bird collection at the British Museum and stays until his death in 1872.
1832, 1850	Edward Lear publishes *Illustrations of the Family of the Psittacidae* in 1832. In 1850 he produces nearly 70 plates for John Gould's *Birds of Europe,* building upon his earlier masterpiece. He turns to writing nonsense verse only after his vision deteriorates in later years.
1833	The word "scientist" is coined, naming collectively the all-male attendees of the third annual meeting for the British Association for the Advancement of Science.[4]
c. 1836	Hokusai, who painted *Snake and Small Bird* (Plate 46), is a mainstream Japanese ukiyo-e artist. He helps popularize Asian art, including Asian bird art, in the West, where much of his art is exported; his ukiyo-e images are less valued in his own country than elsewhere. As early as 1804 he includes a scope in one of his paintings.
1838	Louis Jacques Mande Daguerre, after collaborating with Joseph Nicéphore Niépce until Niépce's death in 1833, develops daguerreotype photography, which gives the first stable image requiring less than thirty minutes of light exposure.
1841	Oil paint in tubes allows artists to move out of the studio and paint outdoors.
	John Cassin builds the bird collection at the Academy of Natural Sciences in Philadelphia into the largest one in the world, with around 29,000 specimens, of which he names 193. Cassin trains no replacement, so his collection goes largely unused. Cassin's primary publication, *Illustrations of the Birds of California, Texas, Oregon, British and Russian America,* though never completed, is the first comprehensive book on birds of the western United States.
1842	Hugh E. Strickland, a geologist with a strong interest in ornithology, creates Strickland's Code setting rules for naming birds; it has the unintended later effect of creating resistance to trinomialism (naming subspecies). In 1848, Strickland writes *The Dodo and Its Kindred,* which correctly identifies the Dodo as a flightless pigeon.
1845	The first use of the term "carrying capacity" in an environmental sense is in a report by the U.S. secretary of state to the Senate.

1845–1850	Phillip Franz von Siebold, former personal physician to King William I, becomes a surgeon in the Dutch colonial army. In Japan he collects specimens until his expulsion in 1828 (on suspicion of spying because he made maps). Back in Leiden he befriends Herman Schlegel, with whom he collaborates on *Fauna japonica,* which includes 175 native birds (described by Temminck), primarily from Nagasaki; it is published between 1845 and 1850. Further work on Japanese birds will be done by Cassin, Richard Bowdler Sharpe, and especially Henry Seebohm, as well as Okada Nobutoshi and Ijima Isao. Ijima will be the first president of the Ornithological Society of Japan, founded in 1912.
1850	James Erxleben, a scientific illustrator who specializes in lithographs of skeletons, works closely with Sir Richard Owen. Owen, the leading comparative anatomist of his era, who coins the word "dinosaur," is involved in establishing the Natural History Museum in London.
1850, 1857	Charles Bonaparte, an ornithologist, publishes the first of four volumes of *American Ornithology.* This supplement to Alexander Wilson's book of the same name is illustrated by Titian Peale and is widely used. Bonaparte also produces *Conspectus generum avium* (1850, 1857), considered the best work of its kind since Latham's *General Synopsis of Birds.*
1850–1858	Gustav Hartlaub, a German ornithologist, publishes *System der Ornithologie Westafrica's* and, in 1857, revises it. In 1852 he establishes the *Journal für Ornithologie,* which remains the leading German bird journal today. He also publishes *Archiv für Naturgeschichte,* in which he urges ornithologists to be wary of evolutionists and avoid anything that weakens the stability of the concept of species.
1854	Ignatio Porro invents the modern "reversing" prism binoculars.
	Henry David Thoreau publishes *Walden.*
1858	Herman Schlegel, an ornithologist, after 33 years at the Rijksmuseum in Leiden, succeeds Coenraad Jacob Temminck as director in 1858. Schlegel collaborates with Phillip Franz von Siebold on *Fauna japonica* (1845–1850), and he eventually creates trinomial nomenclature to identify local varieties, but he rejects Darwin's ideas.
	Philip Lutley Sclater, a British ornithologist, studies birds for half a century and publishes more than 1,300 articles. A lawyer encouraged by Hugh E. Strickland to study natural history, he will specialize in neotropical speciation and zoogeography, strengthening links between Continental and British ornithologists.
	The British Ornithologists' Union is founded.
	Alfred Russel Wallace, a British naturalist, begins collecting exotic insects with Henry Walter Bates and continues collecting in South America and the Malay Archipelago, building a collection that includes 8,000 bird skins. Wallace's paper "On the Tendency of Varieties to Depart Indefinitely from the Original Type" is delivered to the Linnaean Society in England along with Darwin's paper on the subject. Wallace also writes on animal distribution, the origin of species, convergent evolution, and the uniqueness of island species that arise from vanished ancestors.

1859	Charles Darwin, a British naturalist, publishes *On the Origin of Species by Means of Natural Selection, or The Preservation of Favoured Races in the Struggle for Life* (1859), which provides a framework for biological thought by introducing the idea of natural selection (differential reproductive success) as the driving force behind evolution. It was during his five-year expedition around the world on the HMS *Beagle* (1831–1836) that he visited the Galápagos Islands and observed differences in structure and behavior in birds from different islands—observations that proved pivotal in his later studies of selection and the development of modern evolutionary theory.
1860	John Gould, a British naturalist and artist, is consulted by other scientists (e.g., Darwin) on bird matters. His superb images depict around 300 species, most new to science. By 1860 he has published 2,600 lithographs.
1861	A fossil of *Archaeopteryx*, linking reptiles and birds, is discovered in Germany.
1861–1870	Otto Finsch, a wealthy German physician who decides in 1840 to specialize in exotic ornithology, works under Herman Schlegel of the Rijksmuseum in 1861. Finsch publishes a two-volume book on parrots in 1867 and 1868 and works with Hartlaub on two books about birds of the Pacific islands and Africa, *Beitrag zu Fauna Westpolynesiens* and *Die Vögel Ostafrikas*, which are published in 1867 and 1870.
1862	Joseph Wolf, a German naturalist and bird portraitist, considered one of the best bird artists of all time, depicts birds with drama and in relation to their habitat. His work leads ornithologists to switch from using amateur illustrators to professionals.
1863–c. 1900	Ornithological studies in China expand: Robert Swinhoe, an ornithologist, publishes *Checklist of the Birds of China* in 1863 and revises it in 1871. Armand David, a school administrator and traveler, publishes the two-volume *Les oiseaux de Chine*, written with Émile Oustalet, in 1877. The study of birds in China using the Western scientific method is also fostered in the 1870s by Russian interest, especially by the writing of Nikolai Mikhaylovich Przevalski and by his numerous collecting expeditions in Central and East Asia over the decades.
1864	In the United States, George Perkins Marsh publishes *Man and Nature, or Physical Geography as Modified by Human Action*, outlining humanity's destructive impact on the natural environment.
1866	In Germany, Ernst Heinrich Philipp August Haeckel coins the term "ecology" (German: *Oekologie*).
	Édouard Manet, a French impressionist whose best-known depictions of birds include ravens, swallows, and a Grey African Parrot, paints *Young Lady in 1866* (Plate 66).
	Joseph Smit, after painting, lithographing, and hand-coloring 100 images for Philip Lutley Sclater's *Exotic Ornithology* (1866–1869), emigrates from the Netherlands with his family (and the artist John Gerard Keuleman) to Britain. He collaborates with Joseph Wolf and eventually produces 300 plates in 15 major bird books.
1868	An assembly of German farmers appeals to the Austrian and Hungarian foreign minister for an international agreement to protect agriculturally beneficial birds.
1869	The Seabirds Protection Act is passed in the United Kingdom.

1872	Elliot Coues, an American ornithologist and army surgeon who dedicates himself to systematics, names groups of birds that differ in different geographic areas and proposes adding "var." for subspecies varieties. The use of "var." lasts about a decade—until Robert Ridgway switches to trinomial nomenclature for subspecies in 1881. Coues publishes *Key to North American Birds* in 1872, which makes technical information about birds available to nonspecialists. He cofounds the American Ornithologists' Union (AOU) in 1883. As Louis Agassiz Fuertes' uncle, he fosters his nephew's entry into bird art (see Plate 50).
	Robert Angus Smith, a Scottish chemist, coins the term "acid rain" in his book *Air and Rain*.
	The first national park (Yellowstone) is created in the United States.
1873	Felix Bracquemond, a French artist and a friend of Manet's, produces a number of bird pictures, including one depicting efforts of the Acclimatization Society, an organization established to enrich the local flora and fauna, especially where colonialists were nostalgic for familiar species. The establishment of alien bird species, like the European Starling (*Sturnus vulgaris*) in North America, would prove catastrophic.
1874	Spencer Fullerton Baird, an American ornithologist and ichthyologist, is one of the first North Americans to back Darwin. He becomes assistant secretary and later secretary of the Smithsonian Institution in Washington, DC. During his tenure there, which lasts from 1850 to his death in 1887, he expands the collection, having, for example, military servicemen accompanying railroad exploration teams acquire specimens. Upon his death, the Smithsonian collection contains more than 2.5 million natural history specimens. Baird cowrites *A History of North American Birds* with Thomas Brewer and Robert Ridgway, which is published in 1874.
1874–1898	Richard Bowdler Sharpe, a British ornithologist, succeeds George Robert Gray as curator of birds at the British Museum and publishes *Catalogue of the Birds in the British Museum*, a 27-volume enterprise designed to cover birds around the globe as well as all synonyms, literature, and information on their plumage and geographic distribution. He expands the collection from 30,000 to 230,000 specimens, making it the world's largest.
1879	Thomas Henry Huxley, physician, zoologist, and grandfather of Julian and Aldous, serves from 1846 to 1850 as assistant surgeon aboard the SS *Rattlesnake* on its surveying voyage to New Guinea and Australia, where he learns marine biology. Life as an academic follows, and he publishes "On the Classification of Birds" in 1879. A leader in various British scholarly societies, he defends Darwin.
1883	The American Ornithologists' Union (AOU) is founded, as is the Bombay Natural History Society.
1885–1922	Conservation organizations are established, including, for example, the Royal Society for the Protection of Birds, the Royal Australasian Ornithologists Union, the National Audubon Society, and Bird Life International (BLI).
1887	Henry Seebohm, a well-traveled British steel manufacturer and collector who donates his bird skins and eggs to the Natural History Museum in London, writes treatises on Siberian birds in which he analyzes isolation and contact zones as they affect variation in species. Seebohm staunchly defends American trinomialism (the designation of subspecies) in his book *The Geographical Distribution of the Familly Charadriidae* (1887).

1887	Joel Asaph Allen, an ornithologist and the first curator of birds and mammals at the American Museum of Natural History in New York, publishes studies on geographical variation. Allen serves as the first president of the American Ornithologists' Union and the first editor of its journal, *The Auk*. Allen believes that environmental influences can give rise to inherited differences, a theory now known to be incorrect. He devises Allen's Rule, a standard in today's biogeography, that populations of birds and mammals in cooler climates have shorter extremities and therefore can conserve heat.
1888	George Eastman patents the Kodak roll-film camera.
1892	Ernst Hartert, an ornithologist, becomes curator of Lord Walter Rothschild's bird collection in Tring, England, which becomes the second-largest collection in the world and is then sold to the American Museum of Natural History in New York. Hartert describes many new species and from 1903 to 1922 publishes his major work, *Die Vögel der paläarktishen Fauna, 1903–1922*, in which he uses trinomialism (indicating subspecies).
1894	The first high-quality modern binoculars are sold.
	Thomas Edison invents the motion picture.
1899	Christian Mortensen of Denmark institutes a method of numbering and recording leg rings (bands applied to track individual birds).
1900	
1900	The first Christmas Bird Count is held in the United States.
1901	Leon J. Cole, in an address at the Michigan Academy of Science, March 28–30, 1901, recommends banding birds as a method of studying bird migration, basing his suggestion on the use of tags by U.S. Fish Commission to track fish, and publishes a report, "Suggestions for a Method of Studying the Migrations of Birds," in the academy's periodical.
1901–1919	Robert Ridgway, an ornithologist, is named (by Baird) curator of birds at the Smithsonian Institution in 1880, six years after the publication of *A History of North American Birds*, which he cowrote with Baird and Thomas Brewer. In 1901 he publishes the first volume of the multivolume *Birds of North and Middle America*, a milestone in ornithology. Ridgway, working at a time when naming birds (e.g., recognizing subspecies) is a high priority, also self-publishes a book, *Color Standards and Color Nomenclature* (1912), with 53 plates and 1,115 named colors, and is involved in preparing the American Ornithologists' Union *Checklist of North American Birds*.
1902	Convention for the Protection of Birds Useful to Agriculture is signed by 12 countries. It protects 150 bird species but has little effect, in part because Italy—the country is a critical flyway to and from the Mediterranean—does not sign.
1903	Orville and Wilbur Wright achieve the first mechanically powered, sustained human flight in Kitty Hawk, North Carolina.
	President Theodore Roosevelt creates the first Federal Bird Reservation in the United States, on Pelican Island, Florida.
1904	Joseph Grinnell, an ornithologist, founds the Museum of Vertebrate Zoology at the University of California, Berkeley. He publishes more than 500 scientific articles and is the first to describe, in 1904, the role of geographic isolation in speciation (i.e., that geographical barriers or distance separating two populations are key to their differentiation into two species).

1907	Two French brothers, Auguste and Louis Lumière, patent Autochrome Lumière in 1903, marketing it in 1907. This early process for color photography is not surpassed until the mid-1930s.
1908	Frank M. Chapman becomes curator of the ornithology collection at the American Museum of Natural History in New York. While there, he adds habitat groups to the exhibits, hires staff, including Robert Cushman Murphy, and adds specimens, including the enormous Rothschild collection. He founds *Bird-Lore,* a publication for young people in 1899 that becomes *Audubon* magazine. He became an expert on South American birds—their biogeography, ecology, and life histories—and exposes the downside of the plume trade.
1909	Abbott Thayer, an American painter and ornithologist, studies concealing coloration with his son, Gerald, and teaches Fuertes. He explains that bird coloration depends on how light hits it, so bird bellies, which are in shadow, are lighter, and bird backs, which are in light, are darker. He is instrumental in the development of military camouflage.[5]
	Organized bird banding begins in the United States and Britain, and soon thereafter elsewhere, but it does not become a national effort in each country until the 1920s.
1910	Arthur Cleveland Bent, an amateur ornithologist, edits and writes much of the Smithsonian Institution's 44-volume *Life Histories of North American Birds.*
1912	Bruno Leljifors, an innovative Swedish wildlife artist who excels at depicting anatomy, motion, behavior, landscape, and atmosphere, influences bird art, especially that produced since the 1930s. Leljifors provides the impression of birds—their essence, not their feather count—and often depicts predators and their prey.
1913–1914	Oskar Barnack, a German, devises a very portable camera by reducing the size of the negatives and invents the first 35-millimeter still camera. This camera, unlike its heavy, cumbersome predecessors in use since the daguerreotypes of the 1830s, can easily be taken into the field. It will take 11 years before they are manufactured in quantity.
c. 1914	Louis Agassiz Fuertes, who paints *Brewer's Blackbird* (Plate 50) this year, is an American bird portraitist who depicts birds almost exactly as he sees them. He was mentored by his uncle, the ornithologist Elliot Coues, and was taught by the expert in concealing coloration, Abbott Thayer.
1916–1918, 1936	The 1916 Migratory Bird Treaty calling for an end to the hunting of certain birds, is ratified by Canada in 1917 and by the United States the following year. In 1936 it is extended to Mexico under the U.S.-Mexico Convention for the Protection of Migratory Birds and Game Mammals.
1919	Binocular wide-field eyepieces are introduced.
1920	Broadcast radio begins.
1921	The ban on the importation of plumage in the United Kingdom that was introduced to Parliament in 1908 is passed and will take effect in 1922.
1922	Paul Klee, a German-Swiss expressionist artist, describes himself as "taking a line for a walk." His bird depictions are rarely feathered. In *Twittering Machine* (Plate 41), he captures an auditory event visually.
1923–1940	Constantin Brancusi, a Romanian sculptor who worked in Paris from age 28, tries to depict essence rather than relationships and to capture energy as the antithesis of life. He bases numerous works on birds or their eggs, with 16 examples of the *Bird in Space* sequence, dating from 1923 to 1940, including *Bird in Space* (1923), which emphasizes flight rather than the physical bird.

1927	Charles A. Lindbergh makes the first transatlantic flight.
	Charles Elton, a distinguished Oxford biologist, establishes the new field of animal ecology.
1930s	Max Ernst, a German artist, creates Loplop, a bird that is unlike any other, to represent Ernst himself. He uses Loplop for the rest of his life.
1930	Alexander Wetmore becomes the second ornithologist (after Baird) to serve as secretary of the Smithsonian Institution in Washington, DC. Wetmore publishes the important *Systematic Classification for the Birds of the World* this year.
1931	Neville William Cayley publishes the first affordable field guide to birds in Australia, *What Bird Is That?*
1932	The British Trust for Ornithology is founded.
1934	Roger Tory Peterson, an artist, writer, and editor who paints *Evening Grosbeak* (Plate 54) around 1948, is listed at this date because this is the year he also publishes his first field guide. The guide, after an initial rejection, sold out a first printing of 2,000 copies in a week and went on to sell more than 3 million copies. Peterson, a conservationist, is best known not for his art but for his ability to attract tens of millions of people to bird study as an avocation.
1935	Binoculars with antireflective coatings are produced. The coatings increase light transmission by 50 percent.
	Leopold Godowsky and Leopold Mannes perfect a tricolor chemical process, and the resulting color film is marketed under the name Kodacolor by the Eastman Kodak Company.
c. 1935	George Miksch Sutton, an American bird artist, writer, scientist, and museum worker, paints the *Ivory-billed Woodpecker* later used on the cover of *Science* in June 2005 (Plate 67). He was mentored by Fuertes and went on to mentor many younger ornithologists and artists. During his lifetime he illustrates 15 books, 11 of which he wrote, and more than 250 ornithological articles, including an overview of American ornithology and American bird art, *Progress of American Ornithology, 1883–1933*.
1936	Robert Cushman Murphy, a naturalist, ornithologist, and conservationist who spends his career at the American Museum of Natural History in New York, specializes in seabirds. He publishes *Oceanic Birds of South America* in 1936.
1937	Margaret Morse Nice, an American ornithologist, studies individual Song Sparrows in Ohio using colored leg bands to analyze individual roles in territory selection and maintenance and reproductive success, publishing her results in 1937. Altogether, Nice publishes around 60 articles and an autobiography, *Research Is a Passion for Me*.
	Journal of Wildlife Management is founded.
	The Edward Grey Institute of Field Ornithology is founded at Oxford University.
1937–1938	Ducks Unlimited, a hunting-based conservation organization that conserves, restores, and manages wetlands and associated habitats for North America's waterfowl, is founded in the United States; a year later Ducks Unlimited Canada is founded.
1939	Paul Hermann Müller synthesizes DDT. He is awarded the Nobel Prize in Physiology and Medicine in 1948. The first ban on DDT is imposed in 1970.
1940s–1950s	Léo Paul Samuel Robert, an innovative Swiss artist, paints from a bird's-eye view, capturing proper perspective, and aims to convey the essence of the bird. His illustrations appear in Paul Geroudet's six-volume *La vie des oiseaux*, published between 1942 and 1953.

1941	Alden H. Miller, director of the Museum of Vertebrate Zoology at the University of California, Berkeley, publishes a classic study, "Speciation in the Avian Genus Junco," analyzing variation based on a comparison of 11,776 skins. Miller revises this study in 1963 and later extends the results to his 1970 analysis: "Populations, Species and Evolution."
1942, 1963	Ernst Mayr, a German-born ornithologist who studies island birds (of the Solomon Islands and New Guinea), emigrates to the United States and becomes curator of ornithology at the American Museum of Natural History in New York, then Agassiz Professor of Zoology at Harvard University. Mayr publishes *Systematics and the Origin of Species* (1942) and *Animal Species and Evolution* (1963). In the latter he provides a view on speciation that aligns with Grinnell's and becomes central to contemporary views of evolutionary mechanisms.
1944	Francis Lee Jaques is an American engineer turned landscape and bird painter. Well known for his atmospheric style, he is especially adept at incorporating elements that provide reflections and color. He illustrates ornithological texts and other books.
1947	David Lack, an English ornithologist who helped develop radar during World War II and later uses it in studies of bird migration, publishes *Darwin's Finches* in 1947. It is based on work that he began in the Galápagos in 1938 using ecological and genetic differences to investigate differences in finch species. After World War II, Lack becomes director of the Edward Grey Institute of Field Ornithology at Oxford University, where he remains for the rest of his life. He is also credited as the first ornithologist to recognize that banding (ringing) recoveries estimate natural year-by-year mortality rates in wild birds. In 1954 he publishes *The Natural Regulation of Animal Numbers*, and in 1966, *Population Studies of Birds*.
	Peter Scott, an artist and advocate involved in wildlife films, founds Waterfowl Trust (now Wildfowl and Wetlands Trust) in the United Kingdom.
	The Canada Wildlife Service (formerly Dominion Wildlife Service) is founded.
1947–1957	Don Eckelberry, a highly productive bird artist and writer, illustrates Audubon field guides during this decade. He will be the last to paint the Ivory-billed Woodpecker (*Campephilus principalis*) in the wild (see Room 10).
1948	World Conservation Union or International Union for Conservation of Nature and Natural Resources (IUCN), dedicated to natural resource conservation, is founded in Gland, Switzerland.
1950s	Radio telemetry is developed as a tool in wildlife management. It is used in ecology studies at first to determine positional information and later to relay data on an individual animal's physiology and behavior. Over time, the transmitters shrink in size, becoming small enough to glue to small birds, so receiver-equipped cars can track migrants and in-flight conditions.[6]
1950	The International Convention for the Protection of Birds, which extends protection to birds at risk during migration and breeding, is adopted and signed by Austria, Belgium, Bulgaria, Spain, France, Greece, Monaco, the Netherlands, Portugal, Sweden, Switzerland, and Turkey, but has little effect.
	The Nature Conservancy is founded in the United Kingdom and later becomes a leading land-acquisition organization and steward of threatened ecosystems in the United States.
1951	Erwin Stresemann, a German ornithologist, publishes *Entwicklung der Ornithologie*, the first comprehensive history of ornithology, shortly after the end of World War II. In 1974 it is translated into English and given a new section on American ornithology, written by Ernst Mayr.

1951–1953	Niko (Nikolaas) Tinbergen, a Dutch biologist, cofounds with Konrad Lorenz the field of ethology, the biological study of animal behavior. The ethologists examine the function and biological advantage of particular behaviors, devise experiments to test social contexts for those behaviors, and explore their evolutionary significance. The two share the Nobel Prize in Medicine in 1973 with Karl von Frisch, who investigated the communication signals used in social bees. Tinbergen publishes *The Herring Gull's World: A Study of the Social Behavior of Birds* in 1960.
1952	Konrad Lorenz, an Austrian physician and zoologist, further popularizes ethology through *King Solomon's Ring* (1952), a report on the fascinating behaviors of his favorite birds, especially Jackdaws (*Corvus monedula*) and Greylag Geese (*Anser anser*). His explorations have led to an understanding of imprinting and how young animals form social attachments generally, the importance of instinct, and the inheritance of social behavior.
1953	René Magritte, a Belgian surrealist painter, sometimes shows birds evolving from clouds or plants. In *The Enchanted Domain I (Le Domaine Enchanté I)* (Plate 30), he portrays pigeon-plant coevolution.
	The American and British scientists James Watson and Francis Crick describe the double helical structure of deoxyribonucleic acid, or DNA. Crick moved from physics to chemistry and biology in order to investigate the line "between the living and the nonliving." Watson, having studied ornithology, then viruses, turned to the structure of DNA.
1954	Protection of Birds Act in the United Kingdom bans the collection of bird eggs.
1957	Charley Harper is a minimal realist artist whose serigraph *Eskimo Curlew* (Plate 60), like his many other bird images, shows how much we can learn about a bird in an unassuming composition.
1957, 1960, 1961	The first satellites are launched: the first communication satellites (e.g., Sputnik 1, launched in the Soviet Union in 1957), the first weather satellites (e.g., TIROS, for Television InfraRed Observational Satellite, launched in the United States in 1960); and the first manned space vehicles (e.g., Vostok-1, launched in the Soviet Union, and Mercury 3, launched in the United States, both in 1961).
1958	Robert MacArthur publishes in 1958 a classic study showing how very similar species of American wood warblers divide environmental resources in a way that avoids interspecies competition. His application of mathematical theory and his careful analysis of behavior are an important link between behavior and ecology.
1959	Commercial jet travel is available.
1960	M. C. Escher, a Dutch artist, is a master of visual ambiguity: he provides flat planes with depth, makes figure and ground interchangeable, and depicts water flowing uphill. *Ascending and Descending* (Plate 45) illustrates a product of collaborating scientists and artists.
1961	The World Wildlife Fund is registered as a charitable trust in Morges, Switzerland.
1962	The American biologist Rachel Carson publishes *Silent Spring*, identifying the threat from the misuse of pesticides as a springtime without birdsong.
1963	The World Conservation Union (IUCN) writes the Convention on International Trade in Endangered Species of Wild Fauna and Flora (CITES).
1964	The first acrylic paint is available.
1965	John Tuzo Wilson, a Canadian geologist, coins the term "plate tectonics," which revolutionizes the study of geology by showing how Earth's rigid crust is divided into moving plates.

1966	The North American Breeding Bird Survey is initiated.
	National Wildlife Refuge System is initiated in the United States.
1967	Robert MacArthur and E. O. Wilson coin the term "island biogeography" as a result of their studies of how populations and their diversity are achieved in Caribbean islands through a balance between immigration and loss. The concept provides a model for analyzing loss of biodiversity in isolated forests, etc.
	Norman Rockwell, who paints *Woman Observing Bird (Willie Was Different)* (Plate 49) this year, provides covers for *Saturday Evening Post* and other popular images that give him the largest audience of any living artist in history.
1968	Paul Ehrlich publishes *The Population Bomb*, identifying the environmental costs of unfettered population growth.
1969	The American Birding Association is founded.
1970	The Benelux Convention on the Hunting and the Protection of Birds is adopted and signed by three countries (Belgium, Luxembourg, and the Netherlands).
1970–1980	The Club of Rome publishes *Limits to Growth* in 1972, and many laws are enacted to protect the environment. Examples in the United States include the Clean Air Act, the Marine Protection, Research, and Sanctuaries Act, the Clean Water Act, and the Soil and Water Conservation Act.
1971	Greenpeace, an international environmental organization, is founded in Vancouver, Canada.
1973	The U.S. Endangered Species Act is passed into law. By 2007 the law protects 75 endangered and 14 threatened birds.
1978–1981	Personal computers change the way we work and the way we store and access public information, making it far easier to access research on birds and conservation efforts on their behalf, as well as examples of bird photography and bird art.
1979	The European Community (EC) adopts Council Directive 79/409/EEC on the conservation of wild birds, known as the Birds Directive, in response to the 1979 Bern Convention on the conservation of European habitats and species. It protects all naturally occurring birds within the EC (except for Greenland, which was considered anomalous).
1981–1992	As birding, bird photography, and bird art rise in popularity, the United Kingdom passes a law to protect wild birds—Section 1 of the Wildlife and Countryside Act (1981). Partners-In-Flight, a U.S. organization that co-opts citizen birding enthusiasts to gather data for national studies, is launched (1992).
1980s–early 2000s	The threat of global climate change gradually gains international acceptance.
1984–1990	The Canon Company demonstrates the first digital electronic still camera in 1984. Digital imagery and the Internet make their appearance in 1990.
1986–1990	Charles Sibley, Jon Ahlquist, and later Burt Monroe use DNA hybridization to determine genetic similarities between species and restructure the avian family tree.
1987	Bird DNA "fingerprinting" in studies based on the work of Terry Burke and Michael W. Bruford, among others, contributes to our understanding of the sociobiology, demography, and ecology of wild birds.
1988	The World Meteorological Organization (WMO) and the United Nations Environment Program (UNEP) establish the Intergovernmental Panel on Climate Change (IPCC) to evaluate human-induced climate change.
1992	The Habitats Directive extends the protective measures in the Birds Directive of 1979.[7]

1992	The United Nations conference The Earth Summit is held in Rio de Janeiro. William Rees, in Canada, coins the term "ecological footprint."
1997	The Kyoto Protocol limiting the emissions of carbon dioxide and five other greenhouse gases is negotiated in Kyoto, Japan.
	A study by C. P. Chamberlain, J. D. Blum, R. T. Holmes, Xiahong Feng, T. W. Sherry, and G. R. Graves, who sampled feathers from nine sites within the temperate breeding range of the Black-throated Blue Warbler (*Dendroica caerulescens*) in eastern North America, show isotope signatures in feathers that can identify the origin of migratory birds. This is because juvenile birds grow all of their feathers in their natal territories, and prior to migration the blood supply for the now-grown feathers is cut off, making them metabolically inert and traceable back to their place of origin.
1997–2007	There is rising concern about bird viruses. H5N1 appears in 1997; the West Nile virus reaches the United States in 1999; the bird-human transmission of H5N1 is established in 2003; and H5N1 continues to spread, reaching 44 countries by September 2007.
2000	
2006	Computer scientists demonstrate a digital "light field" or "plenoptic" camera with a microlens that is an array of almost 90,000 miniature lenses (like the compound eye of an insect). It can produce photographs that are in focus at all levels of depth.
2007	Global warming models and evidence on especially strong effects in Arctic regions have convinced scientists and most of the public that global warming is a serious problem for natural environments and ecosystems. As the normally ice-blocked Northwest Passage opens for the first time in recorded history, and discussion is under way to add the Polar Bear (*Ursus maritimus*) to the list of threatened and endangered species, studies to record the effects of warming on birds, especially penguins and other seabirds, as well as long-distance migrants, rapidly expand.
	Former U.S. vice president Al Gore shares the Nobel Peace Prize with the Intergovernmental Panel on Climate Change for providing policy makers and the public with evidence for the link between increases in greenhouse gases and global climate change.

APPENDIX 2 A Science Art Checklist for Practitioners

Just as Science Art joins together science and art, the display of Science Art requires textual and visual material. Those who work with Science Art—artists, curators, publishers and other professionals, nature enthusiasts, donors, and patrons—have nontraditional information and extra-routine tasks to consider. The following checklist for labeling Science Art, forming collaborations, exhibiting Science Art, and using Science Art in teaching about environmental issues and in pursuing community activities, as well as for finding Science Art, finding artists who produce Science Art, and supporting Science Artists, serves as a guide.

Labeling Science Art

Scientific Information to Add to Captions
Species portrayed (common name and latinized name)
Aspects of science portrayed
Habitat type (including specific location, if possible)

Personal Information to Add to Long Captions or Catalogue Listings
Artist's biography—see, for example, Carel Pieter Brest van Kempen's biography with Plate 45.
Bibliography, including publications that feature the artist and the artist's work—see, for example, the following selected bibliography for Carel Pieter Brest van Kempen (see Plate 45), prepared in a style widely used in the sciences but here arranged with the most recent publication first. For another style of citation, also widely used, see the bibliography of this book. There are numerous other publications that Brest van Kempen illustrated or that profile his work.

Brest van Kempen, C. P. 2006. *Rigor Vitae: Life Unyielding; The Art of Carel Pieter Brest van Kempen*. Eagle Mountain, UT: Eagle Mountain Publishing.

Beck, D. 2005. *Biology of the Gila Monsters and Beaded Lizards*. Illustrated by C. P. Brest van Kempen. Berkeley: University of California Press.

Pursley, J. M. 2001. *Wildlife Art: 60 Contemporary Masters and Their Work*. New York: Portfolio Press.

Bergman, J. L. 2000. Artists to watch in the new millennium. *Wildlife Art* (January): 28–34.

Smith, C., ed. 2000. *Encyclopedia of Living Artists*. 12th ed. Nevada City, CA: ArtNetwork.

Bestul, S. 1997. Nature's diversity: The art of Carel Pieter Brest van Kempen. *InformArt* (Fall): 34–37.

Tags to Add to Narrow Internet Search Results for Posted Images
See Room 9 and the accompanying notes for more on the difficulties of finding Science Art in Internet searches.

Copyright lines in captions, exhibit lists, and catalogues. In print, insert key words after the artist's name. Online, enter the copyright line as text, not a graphic, so it can be searched.[1] Here are examples.
© 2008 Jane Buck, Science Art—Birds
© 2008 John Doe, Example of Science Art—Birds
John Doe, *Avian Adaptation to Heat*, 2008. Watercolor, 2 × 4 ft (60 × 120 cm), Example of Science Art—Birds

Web page metatags. Insert key words into the metatag of the index html or into the metatag for the specific Web page featuring the image—for example, © 2008 Jane Buck Science Art Birds Nature.[2]

Collaborations

Joint ventures among researchers, artists, and donors make it possible to produce images that might be difficult to market but would be an asset in teaching and public education, especially when portraying aspects of conservation.

Exhibiting Science Art

Physical Venues

To gain access to an exhibit space, whether academic, professional, or commercial, approach the individual responsible for managing it with samples of the work and your credentials. If you lack ties with the sponsor of the space, preferably make the approach in the company of an advocate who does have official ties. Leave a list of who will benefit and why.

Museums and galleries. Request that the artwork be identified as Science Art and that the label and associated information be included in related text, as described above.

Academic venues. Universities have an array of exhibit options, including art spaces, display cases, and miles of walls. High schools, public libraries, medical centers, and so forth, have display cases and wall spaces as well. When producing related Web pages that afford artists and sponsors opportunities to present exhibits virtually, provide printable captions and link related exhibits.

Professional meetings. Annual meetings of nature-related organizations can sponsor exhibits and provide a Web presence.

Other public and private venues. These include, for example, nature centers, offices, lobbies, and corporate-sponsored displays in public buildings Here, again, when the venues have related Web pages, provide printable captions and link related exhibits.

Bare Walls Converted into Exhibit Spaces

Anticipate initial resistance to the proposed conversion of well-positioned bare walls, but wall owners, particularly those concerned about nature and conservation, predisposed to support the arts, and interested in promoting alternative teaching methods, may warm to the idea.

When exhibiting, allow time to navigate permission processes; address security and image conservation issues (including the question of liability and the potential for damage from sunlight); prepare captions, brochures, and publicity; contact potential donors or patrons; and prepare a related online presentation that includes a printable caption.

Create or strengthen links among artists, patrons, and those managing the public spaces—get names and make introductions.

When facing resistance, begin with a suboptimal installation, wait for positive responses to filter back, and reapproach the wall owner.

Internet Venues

The authors have posted a Web page, "Finding Science Art Online," accessible at www.scienceart-nature.org, which includes suggestions and relevant links, as well as information on the Open Directory, where Web sites are categorized by subject and brief descriptions are provided. See also Labeling Science Art, above.

Using Science Art in Teaching about Environmental Issues

Draw attention to environmental issues by increasing artist access to jeopardized habitats and resident species.

Establish art competitions and exhibits featuring jeopardized or restored habitats and resident species.

Encourage collaboration between authors and artists to introduce the public, especially children, to elements of the local natural environment.

Sponsor an online registry of artists specializing in nature images for children to help authors of children's books and Web sites find artists to produce needed images (see Finding Artists Who Produce Science Art, below).

Consult the Environmental Literacy Council Web site, http://www.enviroliteracy.org/, sponsored by the National Science Foundation and the National Endowment for the Arts, to help K–12 students become environmentally literate and acquire the analytical skills necessary to evaluate scientific evidence. See Room 10 and the accompanying notes for more on resources for students and teachers.

Advocate an increase in the visibility of Science Art in higher education coursework—in, for example, classes in ornithology, the history of science, and interdisciplinary courses in environmental science.

Using Science Art in Community Activities

Design public education and community enrichment projects to bring attention to local environmental issues and the efforts undertaken to resolve them.

Design local art competitions to record and exhibit before-and-after images of restoration projects.

Enlist the support of local businesses for community art programs designed to enrich and educate neighborhood residents. Request their sponsorship of science art competitions to defray exhibit costs.

Finding Science Art

Finding Science Art online, in publications, and on exhibit is a rather frustrating experience. For ways to find it online see Internet Venues, above. Also consult art databases, such as ARTstor and AMICO, which may be available through a university's or library's online resources. For more discussion see the Introduction and Room 10. But the best way to find it will be by getting Science Art listed as a category of art:

Recommend the addition of Science Art as a category of art in libraries and reference books.

Recommend the use of Science Art as a category to arts commissions, granting agencies, art groups (such as societies that support wildlife art), major nature-related art competitions and exhibitions, galleries, and museums.

Recommend the use of Science Art as a subject heading in art and image databases such as ARTstor and AMICO, mentioned above.[3]

Finding Artists Who Produce Science Art

Consult online registries for information on artists producing Science Art, as well as publishers of Science Art and organizations, galleries, and museums that support—or might support—Science Art. The Artist Registry for Ornithological Researchers at birds.stanford.edu, for example, is a resource for bird-related images.[4]

Advocate the launching of additional registries that feature Science Art and specialize in particular flora or fauna, such as registries that provide information on artists producing images that are mammal-related, fish-related, or endangered-plant-related.

Support for Science Artists

To commission or exhibit artists who specialize in Science Art consult artist registries specializing in Science Art, nature-related organizations, art groups, galleries, and museums. Examples are the Artist Registry for Ornithological Researchers, regional nature reserves, the Society of Animal Artists, and the Leigh Yawkey Woodson Museum.

To find material of use in preparing Science Art consult Web sites that provide information on the biology of birds and their conservation status. Other Web sites provide images that can verify certain behaviors or particular plumage or other characteristics of the species being portrayed. Yet others provide timelines that can help verify the historical context when aspects of culture will be included in the art. Information on bird photography is available in many books and at the Web sites of high-profile individual photographers—for example, Arthur Morris—which can be very informative. See also the authors' Web site, cited under Internet Venues, above.[5]

Notes

Introduction

1. The calculation is based on a generation time of twenty years.
2. On the UNESCO award see Strosberg, *Art and Science*, 232–233. On the return date see http://www.keo.org/uk/pages/faq.html#q3 (accessed February 4, 2005), which explains that the return in 50,000 years was selected as "the mirror date to a milestone in the evolution of our species: the first traces of Art reveal the human capacity for abstract thought and symbolic expression." Additional information on KEO can be accessed at this address: http://www.keo.org/uk/pages/faq.html#q19 (accessed February 9, 2005).
3. According to the Audubon Society in 2004, "Nearly 60 million birders in North America have made birding the second most popular outdoor activity after gardening, spending billions of dollars annually on birding supplies." http://www.audubon.org/bird/wb.html (accessed October 23, 2004). On the research published in 2001 see artist-registry.stanford.edu (accessed October 21, 2004).

Lower Gallery

1. Key periods related to human and bird relationships:
    Upper Paleolithic: The Upper Paleolithic was cold and dry with a lot of climatic variability, which led to frequent changes in diet for both birds and humans and to advances in technology. It lasts until the introduction of farming at the beginning of the Holocene, 10,000 BCE.
    Ancient Civilizations and the Beginning of the Common Era (CE): Birds are primarily depicted as icons (religious, mythological, or political) or as resources for human use.
    Middle Ages and the Renaissance: Birds are still primarily shown as icons or as resources for human use.
    Early Scientific Revolution: In the West larder art (estate paintings including those that portray hunting trophies) is in its heyday. Pictorial evidence of trade in local and exotic birds expands, and occasional demonstrations of "drawing-room science" are produced. Pictorial evidence of the popularity of cage birds as pets also expands, as does the number of images that reflect aspects of bird biology, like predator-prey relationships, competition, and foraging.
    The Information Age: Some images convey conservation sensitivity (or lack thereof), showing range reduction, the effects of increases in cat populations, species requiring protection, and so forth. Bird prints showing aspects of bird biology become increasingly popular, and there is an increase of pictures of birds in all graphic forms —from illustrations in the ornithological literature, field guides, and commercial media to fine art and photographs. (Former Biology of Birds Web site [site was password protected].)
2. A systematic survey of the birds recorded in the art of ancient Egypt and a list of mummified birds is in Houlihan, *Birds of Ancient Egypt*, v, 46–49, 140.
3. A. Leroi-Gourhan, *Treasures of Prehistoric Art* (New York: Harry N. Abrams, 1967), 502. We thank Paul Hull for bringing to our attention the wealth of Paleolithic stone objects that also portray birds. See, for example, the analysis of French Paleolithic objects found in D. Buisson and G. Pinçon, "Nouvelle lecture d'un galet gravé de Gourdan et essai d'analyse des figurations d'oiseaux dans l'art paléolithique Français," *Antiquités Nationales* 18–19 (1986–1987): 85–89. Buisson and Pinçon list swans, ducks, and other web-footed birds, grouse and other gallinaceous birds, cranes, bustards, stilt and other marsh species, raptors, owls, and perching birds. In an article by Stephen J. Gould, discussion of the owl discounts the possibility that the artwork was forged: the ground beneath the overhang on which it was engraved had collapsed, opening a crater that put the overhang out of reach. Gould, "Rhinos and Lions and Bears (Oh, My!)," *Natural History* (May 1995): 30–34.
4. Here is the official Web site for Chauvet Cave: http://www.culture.gouv.fr/culture/arcnat/chauvet/en/index.html (accessed August 5, 2004).
5. The following sources were helpful in our evaluation: "French Cave Yields Stone Age Art Gallery," *Science News* (January 1995): 28, 52–53; J. de la Torre, *Owls: Their Life and Behavior* (New York: Crown, 1990), 20, 110; Fischman, "Painted Puzzles," 614; Gould, "Lesson from the Old Masters," 16–22, 58–59; Morell, "Stone Age Menagerie," 54–62.

6. For argument's sake, we could say that the stripes represent the heavy vertical markings of the ventral plumage. The ears of both owl species are long (although it is true that those of the Long-eared Owl [Moyen Duc] are held more vertically), and the range of both owl species covers Europe now and probably did 30,000 years ago.

7. Tomorrow's viewers will probably have even more information. Researchers could find more caves painted during the Paleolithic and the bones of more owls in the vicinity, but we may never reach the point of identifying the Chauvet owl with certainty, just as we may never know whether the many painted hand prints with their missing fingers, like those found at Cosquer Cave in large numbers (see Plate 17), represent mutilations, sign language, or even evidence of widespread frostbite, but the menagerie recorded on the walls of Chauvet Cave tell us one thing with certainty that changes our view of our ancestors: 30,000 years ago they could depict entire scenes and events—and not just in silhouette—like the fight recorded between rhinoceroses found not far from the Chauvet owl. E. H. Gombrich, "The Miracle at Chauvet," *New York Review of Books* (November 14, 1996): 9–10.

8. Clayton, *Chronicle of the Pharaohs*, 204–205.

9. Said, *The River Nile*, ix; 128.

10. Janson, *History of Art*, 60–61; M. Saleh and H. Sourouzian, *Official Catalogue: The Egyptian Museum, Cairo*, trans. P. der Manuelian and H. Jacquet-Gordon (Mainz, Germany: Phillip von Zabern, 1987), pl. 31; Wilkinson, *Reading Egyptian Art*, 82–83.

11. W. B. Clark, *Medieval Book of Birds*, xi, 14, 119, 135. The book was written when Hugh of Fouilloy was prior of Saint-Nicolas-de-Regny, an Augustinian house near Amiens. Ingersoll, *Birds in Legend, Fable and Folklore*, 133–135.

12. T. M. Plautus, *Amphitryo, Asinaria, Aulularia, Bacchides, Captivi*, trans. Paul Nixon, fee-free eBook, #16564, release date: August 20, 2005, available from Project Gutenberg (www.gutenberg.net).

13. Here are the full stanzas, but not the whole poem, from J. Gay, "The Farmer's Wife and the Raven," Fable xxxvii, in J. Addison, J. Gay, and W. Somerville, *The Poetical Works of Addison; Gay's Fables; and Somerville's "Chase," with Memoirs and Critical Dissertations by the Rev. George Gilfillan*, fee-free eBook, #10587, release date: January 4, 2004, available from Project Gutenberg (www.gutenberg.net), at: http://www.gutenberg.org/etext/10587.

> Betwixt her swagging panniers' load
> A farmer's wife to market rode,
> And, jogging on, with thoughtful care
> Summed up the profits of her ware;
> When, starting from her silver dream,
> Thus far and wide was heard her scream:
>   "That raven on yon left-hand oak
>   (Curse on his ill-betiding croak)
> Bodes me no good." No more she said,
> When poor blind Ball, with stumbling tread,
> Fell prone; o'erturned the pannier lay,
> And her mashed eggs bestrewed the way.
>   She, sprawling in the yellow road,
> Railed, swore and cursed: "Thou croaking toad,
> A murrain take thy whoreson throat!
> I knew misfortune in the note."
>   "Dame," quoth the raven, "spare your oaths,
> Unclench your fist, and wipe your clothes.
> But why on me those curses thrown?
> Goody, the fault was all your own;
>
> For had you laid this brittle ware,
> On Dun, the old sure-footed mare,
> Though all the ravens of the hundred,
> With croaking had your tongue out-thundered,
> Sure-footed Dun had kept his legs,
> And you, good woman, saved your eggs."

For more on raven lore see T. M. Gehring, "Potential Predatory Attack by Common Ravens on Porcupines," *Wilson Bulletin* 105.3 (1993): 524–525; Ingersoll, *Birds in Legend, Fable and Folklore*, 168–176; Welty and Baptista, *Life of Birds*, 563; Heinrich, *Mind of the Raven*, 194–197.

14. For more on ravens see http://www.pbs.org/lifeofbirds/brain/index.html (accessed August 31, 2007).

15. Uglow, *Lunar Men*, xv; B. Tattersall, *Stubbs and Wedgwood: Unique Alliance between Artist and Potter* (London: Tate Gallery, 1974), 84, 113. Wedgwood notes that Stubbs was analytical and "has been a drawing master amongst other things. . . . He begun teaching perspective to his pupils, which he believes to be just as rational a method in draw.ᵍ as learning the letters first is [in] acquiring the art of reading, & he would have the learner to copy nature & not drawings." For more on the Lunar Society see Plate 44.

16. P. Sims, *Whitney Museum of American Art: Selected Works from the Permanent Collection* (New York: Whitney Museum of American Art in association with W. W. Norton, 1985), 88, 253; D. C. Miller and A. H. Barr, Jr., eds., *American Realists and Magic Realists* (New York: Museum of Modern Art, 1943), 58; www.sjmusart.org/dynamic_content/teacherPDFs/whitney Museum_book.pdf (accessed August 6, 2004).

17. C. Holden, "Inching Toward Movement Ecology," *Science* 313 (2006): 779–782.

18. M. F. Boynton, ed., *Louis Agassiz Fuertes: His Life Briefly Told by His Correspondence* (New York: Oxford University Press, 1956), 212.

19. On Bushmen folklore see http://fraktali.849pm.com/text/archive/afr/bleek.htm (accessed August 10, 2004); W. H. I. Bleek and L. C. Lloyd, *Specimens of Bushman Folklore* (London: George Allen, 1911), 134–145. On ostriches in ancient Egypt see http://www.fsmitha.com/h1/ch02.htm (accessed August 16, 2005). Biblical passages forbidding the eating of ostrich include Leviticus 11:13 and Deuteronomy 14:12. One criticizing the parenting skills of the female ostrich is Job 39:17. On Commodus see Toynbee, *Animals in Roman Life and Art*, 238. On ostrich parenting skills see B. C. R. Bertram, *The Ostrich Communal Nesting System* (Princeton, NJ: Prince-

ton University Press, 1992), 1, 7–12, 14–16, 186–187. On early bird art see L. Pericot-García, J. C. Galloway, and A. Lommel, *Prehistoric and Primitive Art* (New York: Harry N. Abrams, 1967), 120; Pollard, *Birds in Greek Life and Myth*, 86, 106; Klingender, *Animals in Art and Thought*, 72–76; Lysaght, *Book of Birds*, 136; N. Manlius, "The Ostrich in Egypt: Past and Present," *Journal of Biogeography* 28.8 (2001): 945; M. Marini, *Caravaggio: Michelangelo Merisi da Caravaggio "Pictor Praestantissimus": La Tragica Esistenza, la Raffinata Cultura* (Rome: Newton Compton, 1987), 278–279; K. Michalowski, *Art of Ancient Egypt* (New York: Harry N. Abrams, 1969), 400, 454; Toynbee, *Animals in Roman Life and Art*, 237–240.

20. The latinized name for the Ostrich family is Struthionidae.
21. The massive Giant Elephant Bird was 10 feet (3.2 meters) tall. Fuller, *Extinct Birds*, 34.
22. Bertram, *Ostrich Communal Nesting System*, 186.
23. For more information on Egyptian birds and bird art see P. Houlihan, "Bird Life along the Ancient Nile," *Ancient Egypt Magazine* 3:1 (July–August 2002); P. Houlihan, "Birds in Ancient Egypt: The Plumage of the Gods," *Ancient Egypt Magazine* 3:2 (September–October 2002), at http://www.ancientegyptmagazine.com/birds13.htm (accessed September 7, 2004); and http://www.ancientegyptmagazine.com/birds14.htm (accessed September 7, 2004). For a tour of Horemheb's tomb see http://www.osirisnet.net/tombes/pharaons/horemheb/e_horemheb_part1.htm (accessed August 11, 2004). More than one title is associated with the painting. One, "The Chief Fowler Ptah-Mose," is used by Sir Alan H. Gardiner (C. Aldred, *The Development of Ancient Egyptian Art from 3200 to 1315 B.C.* [London: Alec Trianti, 1952], 61); another, "Keeping Pelicans and Collecting Eggs," is used in P. Germond and J. Livet, *An Egyptian Bestiary: Animals in Life and Religion in the Land of the Pharaohs* (London: Thames and Hudson, 2001), fig. 106.
24. The image shown in Plate 12 is not the only Egyptian example of pelicans. A similar scene is recorded in the Sun Temple of Niuserre at Abu Gurob and in the mastaba of *Niankhkhnum and Khnumhotep*, both produced during the Fifth Dynasty (2498–2345 BCE). Houlihan, *Birds of Ancient Egypt*, 10. For more on birds in Egyptian art see http://library.thinkquest.org/C0121761/22.htm (accessed August 11, 2004); Aldred, *Development of Ancient Egyptian Art*, 61; Ehrlich, Dobkin, Wheye, and Pimm, *Birdwatcher's Handbook*, 20; Germond and Livet, *Egyptian Bestiary*, 46 (see plate 47). Houlihan notes that in the fourth century Horapollo wrote that although Egyptians ate pelican meat, priests were to abstain. Houlihan, *Birds of Ancient Egypt*, 10, 13.
25. M. Caygill, *Treasures of the British Museum* (London: British Museum Publications, 1985), 129.
26. For more on Sutton Hoo see http://www.thebritishmuseum.ac.uk/explore/families_and_children/online_tours/sutton_hoo/sutton_hoo.aspx (accessed August 10, 2004); Caygill, *Treasures of the British Museum*, 129–134; A. Caleca, D. Gioseffi, G. L. Mellini, and L. R. Collobi, *British Museum, London*, Great Museums of the World (New York: Newsweek, 1967), 122, 124; Frederick II, *Art of Falconry*, 451.
27. For a brief overview on Boucher see P. Stein, "François Boucher (1703–1770)," in *Timeline of Art History* (New York: Metropolitan Museum of Art, 2000), available at http://www.metmuseum.org/toah/hd/bouc/hd_bouc.htm (October 2003) (accessed September 9, 2007); M. Hyde and M. Ledbury, eds., *Rethinking Boucher* (Los Angeles: Getty Research Institute, 2006). B. M. Stafford notes that this painting was produced when literacy was rising and people were increasingly making the switch from reading aloud to reading silently. Stafford, *Artful Science*, 47, 282. The quotation from Boucher, "trop verte et mal eclairé," is noted in http://museumnetworkuk.org/landscapes/gallery/objectpages/boucher.htm (accessed August 31, 2007).
28. For more on Homing Pigeons see C. E. Jackson, *Bird Painting: The Eighteenth Century* (Woodbridge, Suffolk, England: Antique Collectors' Club, 1994), 28; J. Hansell, *The Pigeon in History, or The Dove's Tale* (Bath, England: Millstream Books, 1998), 55, 132–138; Welty and Baptista, *Life of Birds*, 518–522, 524–525, 537.
29. For more on Sally M. Berner's painting, see http://www.artistsmagazine.com/2001_animal.html (accessed August 10, 2004) and http://www.natureartists.com/artists/artist_biography.asp?ArtistID=974 (accessed August 10, 2004).
30. S. Berner, quoted in *Birds in Art: 2000* (Wausau, WI: Leigh Yawkey Woodson Art Museum, 2000), 17.
31. American Meat Institute Fact Sheet, "Overview of U.S. Meat and Poultry Production and Consumption" (March 2003), available at http://www.meatami.com/Content/NavigationMenu/PressCenter/FactSheets_InfoKits/FactSheetMeatProductionandConsumption.pdf (accessed September 7, 2004).
32. For more on William Lishman's work see http://www.operationmigration.org/ (last accessed July 24, 2006).
33. For more on Cosquer Cave see http://www.showcaves.com/english/fr/caves/Cosquer.html (accessed 2004).
34. American Ornithologists' Union, *Check-List of North American Birds*, 6th ed. (Lawrence, KS: Allen Press, 1983), 242; Guthrie, *Nature of Paleolithic Art*, 299.
35. More fully: "In northern European waters, Great Auks were also hunted. One Icelander expert in capturing them reported that 'the wings are kept close to the sides when the bird is at rest, but a little out (so that light shows under it) when it begins to run. That is, run from humans.'" C. Cokinos, *Hope Is the Thing with Feathers: A Personal Chronicle of Vanished Birds* (New York: J. P. Tarcher/Putnam, 2000), 317. Is the Cosquer auk on land or in the water? We suspect that it is on land. Auks swim like ducks and dive like penguins. The bird in the drawing does not resemble a duck afloat, and the artist would have

had difficulty watching a dive. R. Dale Guthrie sees it differently, however. Guthrie, *Nature of Paleolithic Art,* 447. See also Gombrich, "Miracle at Chauvet," 10.

36. Fuller, *Great Auk,* 65–66, 68. Fuller quotes the observation by J. Allen, originally published in 1876: "The birds were then easily killed, and their feathers removed by immersing the birds in scalding water which was ready at hand in large kettles set for the purpose. The bodies were used as fuel for boiling the water."
37. Clottes and Courtin, "Neptune's Ice Age Gallery," 65; Ruspoli, *Cave of Lascaux,* 28–30.
38. "Huang Quan," in *The Grove Dictionary of Art Online* (Oxford University Press), http://www.groveart.com (accessed August 13, 2004); Z. Jiajin, comp., and G. Hutt, ed., *Treasures of the Forbidden City* (New York: Viking, 1986), 86–87. Cochoas are medium-sized insect- and mollusk-eating birds in the genus Cochoa. They were formerly considered members of the thrush family, Turdidae, but now are more often treated as part of the Old World Flycatcher family, Muscicapidae. An image and additional information on these birds are available at http://encyclopedia.thefreedictionary.com/Cochoa (accessed August 13, 2004); Morell, "Arduous Journeys," 783–784.
39. Lysaght, *Book of Birds,* 38.
40. For more on Pisano see http://gallery.euroweb.hu/html/p/pisano/andrea/omechani.html (accessed August 12, 2004); and A. Bertram, *Florentine Sculpture* (London: Studio Vista, 1969), 17–22. For more on Greek art and architecture see http://encarta.msn.com/text_761561691__1/Greek_Art_and_Architecture.html (accessed August 12, 2004). For more on this image of Daedalus see http://ccat.sas.upenn.edu/~jfarrell/courses/spring96/myth/mar25.html (accessed August 12, 2004).
41. MIT's Tenney L. Davis translated Roger Bacon's description of a flying machine: "It is possible that a device for flying shall be made such that a man sitting in the middle of it and turning a crank shall cause artificial wings to beat the air after the manner of a bird's flight." *Roger Bacon's Letter: Concerning the Marvelous Power of Art and of Nature and Concerning the Nullity of Magic* (Easton, PA: Chemical Publishing Co., 1932), 3. Leonardo left an array of sketches and descriptions. His most complete image is in the Codex Atlanticus (341r), held at Pinacoteca Ambrosiana in Milan, where he shows a rotation system that is activated using the heel and hand so that vertical and horizontal flight are both possible. In 1488, Leonardo drew versions of a flying machine featuring foot pedals.
42. For more on bird flight see Ehrlich, Dobkin, and Wheye, *Birder's Handbook,* 507, 509, 511.
43. The first specimen known to science was discovered in Germany in 1861; the Berlin specimen was discovered in 1877. Feduccia, *Origin and Evolution of Birds,* 111–112. Hind-limb feathers on specimens discovered in China appear to be important for gliding and possibly flying. P. Christiansen and N. Bonde, "Body Plumage in *Archaeopteryx:* A Review, and New Evidence from the Berlin Specimen," *Comptes Rendus Palevol* 3 (2004): 99–118; F. Zhang and Z. Zhou, "Palaeontology: Leg Feathers in an Early Cretaceous Bird," *Nature* 431 (2004): 925; X. Xu, Z. Zhou, X. Wang, X. Kuang, F. Zhang, and X. Du, "Four-Winged Dinosaurs from China," *Nature* 421 (2003): 335–340; N. Longrich, "Structure and Function of Hindlimb Feathers in *Archaeopteryx lithographica,*" *Paleobiology* 32.3 (2006): 417–431.
44. W. E. Wilson, "Balthazar Sage (1740–1824)—The History of Mineral Collecting: 1530–1799)," *Mineralogical Record* 25.6 (November–December 1994): 49, at http://static.highbeam.com/t/themineralogicalrecord/index.html (accessed 2004).
45. W. H. Adams, ed., *The Eye of Thomas Jefferson* (Washington, DC: National Gallery of Art, 1976), 60. For more on Colson see http://europeanpaint.brinkster.net/html/bio.asp?numcol=8 (accessed 2004).
46. L. Stroncek, quoted in *Birds in Art: 1989* (Wausau, WI: Leigh Yawkey Woodson Art Museum, 1989), 115. For information provided in *The Birds of North America Online* see http://bna.birds.cornell.edu/BNA/account/Canada_Goose/MIGRATION.html (accessed October 1, 2006). On folklore linking bird behavior to changes in weather see Ingersoll, *Birds in Legend, Fable and Folklore,* 278. The Ingersoll book includes the following lore on page 222: "The behavior of some species indicates a change in local weather. For example, Acorn Woodpeckers (*Melanerpes formicivorus*) drill holes in trees and cache acorns within them. Rising humidity causes cached acorns to swell and protrude from the holes. For the Gualala Indians of Sonoma, California, an increase in drumming, as the woodpeckers tapped the cached acorns back in place, served as an 'acoustic barometer' forecasting a chance of rain."
47. R. J. Ray, B. R. Cobleigh, M. J. Vachon, and C. St. John, "Flight Test Techniques Used to Evaluate Performance Benefits during Formation Flight," NASA/TP-2002-210730 NASA Dryden Flight Research Center (2002), at http://www.nasa.gov/centers/dryden/pdf/88742main_H-2500.pdf (accessed 2004 and September 1, 2007); J. P. McCarty, "Ecological Consequences of Recent Climate Change," *Journal of the Society for Conservation Biology* 15 (2001): 320, at http://www.blackwell-synergy.com/doi/abs/10.1046/j.1523-1739.2001.015002320.x (accessed September 3, 2007); C. D. MacInnes, E. H. Dunn, D. H. Rusche, F. Cooke, and F. G. Cooch, "Advancement of Goose Nesting Dates in the Hudson Bay Region, 1951–86," *Canadian Field Naturalist* 104 (1990): 295–297. For more on the responses of plants and animals to recent climate change see C. Parmesan, "Ecological and Evolutionary Responses to Recent Climate Change," *Annual Review of Ecology, Evolution, and Systematics* 37 (2006): 637–669, available online at http://chge.med.harvard.edu/programs/education/course_2007/topics/02_14/readings.html#pimm (accessed September 1, 2007); Ehrlich, Dobkin, and Wheye, *Birder's Handbook,* 59; Ehrlich, Dobkin, Wheye, and Pimm, *Birdwatcher's Handbook,* 43; and Thomas

et al., "Extinction Risk from Climate Change," 145–148.

48. Ehrlich, Dobkin, and Wheye, *Birder's Handbook*, 53. The oil comes from a specialized gland above a bird's tail and is used to keep feathers waterproof and supple. Feathers with a quill (contour feathers as opposed to down) have a vane made of parallel barbs that are held together with tiny hooks. When contour feathers are disturbed and the tiny hooks come undone, a swipe through the bill zips them back into place.

49. Ruspoli, *Caves of Lascaux*, 73; Guthrie, *Nature of Paleolithic Art*, 280, 279.

50. For a discussion on dating rock art, which includes comments on the interpretations of Breuil, Leroi-Gourhan, and others, see Christian Züchner's overview at http://www.uf.uni-erlangen.de/zuechner/rockart/dating/archdating.html (accessed September 9, 2007).

51. S. J. Gould, "Up Against a Wall," *Natural History* (July 1996): 16–22, 70–73; Gould, "A Lesson from the Old Masters," 16–22, 58–59; H. Bégouën and L'Abbé H. Breuil, *Les Cavernes du Volp* (Paris: Arts et Métiers Graphiques, 1958), 17–18; Guthrie, *Nature of Paleolithic Art*, 224.

52. L. Kurke, *Coins, Bodies, Games, and Gold: The Politics of Meaning in Archaic Greece* (Princeton, NJ: Princeton University Press, 1999), 4–9; Pollard, *Birds in Greek Life and Myth*, 141–142; C. Seltman, *A Book of Greek Coins* (London: Penguin Books, 1952), 11; C. Seltman, *Greek Coins* (London: Thames and Hudson, 1966), 296–297.

53. Ehrlich, Dobkin, and Wheye, *Birder's Handbook*, 639.

54. A seer in the play described the "happy omen" when Greek troops under two kings marched off to fight the Trojans:

> They got a happy omen—two eagles,
> kings of birds, appeared before the kings of ships.
> One bird was black, the other's tail was white,
> here, close to the palace, on the right,
> in a place where everyone could see.
> The eagles were gorging themselves,
> devouring a pregnant hare
> and all its unborn offspring,
> struggling in their death throes still.
> . . .
> Then the army's prophet, Calchas,
> observing the twin purposes
> in the two warlike sons of Atreus,
> saw the twin leaders of the army
> in those birds devouring the hare.
> He then interpreted the omen, saying,
> "In due course this expedition
> will capture Priam's city, Troy."

The play is available in book form, of course, and at http://www.mala.bc.ca/~johnstoi/aeschylus/aeschylus_agamemnon.htm (accessed August 17, 2004).

55. For more on the lore see W. B. Clark, *Medieval Book of Birds*, 83, 85. We note here Giraldus Cambrensis's description of Barnacle Geese in his *The Topography of Ireland*, trans. T. Forester, ed. T. Wright (Cambridge, Ontario, Canada: In Parentheses Publications, 2000), 20-21, available online at http://onlinebooks.library.upenn.edu/webbin/book/lookupid?key=olbp35530: "They resemble the marsh-geese, but are smaller. Being at first gummy excrescences from pine-beams floating on the waters, and then enclosed in shells to secure their free growth, they hang by their beaks, like seaweeds attached to the timber. . . . I have often seen with my own eyes more than a thousand minute embryos of birds of this species on the seashore, hanging from one piece of timber covered with shells, and already formed. No eggs are laid by these birds after copulation, as is the case with birds in general; the hen never sits on eggs in order to hatch them; in no corner of the world are they seen either to pair, or build nests. Hence, in some parts of Ireland, bishops and men of religion make no scruple of eating these birds on fasting days, as not being flesh, because they are not born of flesh." Additional information on manuscripts and art from this period is found in Klingender, *Animals in Art and Thought*, 353.

56. S. K. Pierce and T. K. Maugel, *Illustrated Invertebrate Anatomy: A Laboratory Guide* (New York: Oxford University Press, 1987), 238.

57. Farid ud-Din Attar, *The Conference of the Birds (Manteq at-Tair)*, trans. A. Darbandi and D. Davis (New York: Penguin Books, 1177/1984), 16–17, 20. Chaucer's poem and an English translation are available online from the Chair of Medieval English Literature and Historical Linguistics at Düsseldorf University, at http://www.phil-fak.uni-duesseldorf.de/anglist1/html/parlement_of_foulys.html (accessed September 3, 2007).

58. For a short commentary from Düsseldorf University see http://web.phil-fak.uni-duesseldorf.de/~holteir/companion/Navigation/Authors/Chaucer/WorksChaucer/PF/pf.html (accessed August 19, 2004). For information on leks see Ehrlich, Dobkin, and Wheye, *Birder's Handbook*, 259.

59. Jackson, *Bird Painting: The Eighteenth Century*, xxx, 41–44. Christine Jackson has listed the birds in the painting as follows: "Hoopoe, Barn Owl, Great Grey Shrike, Brambling, Kestrel, Jay, Black Woodpecker, Bullfinch, Song Thrush, Nutcracker, Osprey, Blue-and-Gold Macaw, Scarlet Macaw, Woodpecker, Long-eared Owl, Magpie, White Stork, Eagle, Kite, Bluethroat, Goldfinch, Hawfinch, Crested Tit, Robin, Waxwing, Smew, Red-footed Falcon, Hobby, Starling, Great Tit, Wallcreeper, Purple Heron, Cassowary, Crowned Crane, Green Woodpecker, Ostrich, Demoiselle Crane, Crane, Flamingo, Goose, Turkey, Bustard, Muscovy Duck, Pheasant, Rock Partridge, Golden Eagle (the President), Little Bustard, Wagtail, Kingfisher, Oystercatcher, Squacco Heron, Black-necked Grebe, Guineafowl, Lapwing, Great Crested Grebe[s], Bittern, immature gull, Teal, Shoveller, Garganey, Pochard, Goosander, Goldeneye, Mute Swan, Pelican, Mallard, Coot[s], Grey Heron, Peacock, Cormorant." De Hamilton's assembly includes a number of dupli-

cates and exotics, including the two mentioned in the poem—peacock and parrot (popinjay), which are represented by Scarlet and Blue-and-Gold Macaws—and three that are not mentioned (Red-faced Lovebird, Cassowary, and Crowned Crane). Open bills indicate vocalizing. Note the paucity of pairs.

60. For information on Magritte from the Tate Collection's Web site see http://www.tate.org.uk/servlet/ArtistWorks?cgroupid=999999961&artistid=1553&tabview=bio (accessed 2004).

61. P. R. Ehrlich and P. H. Raven, "Butterflies and Plants: A Study in Coevolution," *Evolution* 18 (1964): 586–608.

62. Jackson, *Bird Paintings: The Eighteenth Century*, 27.

63. In the mid-1990s, U.S. residents spent an estimated $5–$6 billion annually to watch birds. This compares with the $5.8 billion they spent on movies and the $5.9 billion on sports events.

64. Herodotus, *The Histories*, trans. Aubrey de Sélincourt (Baltimore, MD: Penguin Books, 1954), 128.

65. Ibid., 127–128.

66. Pimm et al., "Human Impacts on Bird Extinctions."

67. A. Feduccia, ed., *Catesby's Birds of Colonial America* (Chapel Hill: University of North Carolina Press, 1985), 62. This estimate of the Passenger Pigeon population was made by A. W. Schorger. John James Audubon took 350 live (market-bought) birds to England in 1830. The last "official" sighting of a wild Passenger Pigeon was in 1900.

68. Ruspoli, *Cave of Lascaux*, 191–192; Read, *Icon and Idea*, 56.

69. Heinrich, *Mind of the Raven*, 145, 242–243; Guthrie, *Nature of Paleolithic Art*, 162; Marzluff and Angell, *In the Company of Crows and Ravens*, 2–3.

70. Ruspoli, *Cave of Lascaux*, 191–192. Readers who take a tour, either online or in France at Lascaux II, might take a look at the nearby rhinoceros and consider whether it might have been part of the scene and whether the bird is surmounting a staff or spear thrower. Also worth noting are the bird's long "legs," the man's short arms, and his hands, which are short two fingers.

71. Ibid., 72–73; J. Diamond, "New Guineans and Their Natural World," in R. Kellert and E. O. Wilson, eds., *The Biophilia Hypothesis* (Washington, DC: Island Press, 1993), 251, 267–268.

72. For a BBC report on mummified animals in ancient Egypt see http://news.bbc.co.uk/1/hi/world/middle_east/1254835.stm (accessed August 22, 2004); Houlihan, *Birds of Ancient Egypt*, 48.

73. Thoth was also an ape and a cat and might also be shown as an ibis-headed man beneath a crescent moon. K. Clark, *Animals and Men*, 75. For more information on the Sacred Animal Necropolis at North Saqqara, see J. Gosling, P. Manti, and P. T. Nicholson, "Discovery and Conservation of a Hoard of Votive Bronzes from the Sacred Animal Necropolis at North Saqqara," *Egyptology* 2.1 (2004), which can be ordered at http://www.palarch.nl/Archaeology/archive.htm (accessed September 1, 2007); J. D. Ray, "The World of North Saqqara," *World Archaeology* 10.2 (1978): 149–157; additional coverage of the archaeological sites can be found at http://www.egyptsites.co.uk/lower/saqqara/cemeteries/cemeteries.html (accessed August 22, 2004); a regional map can be downloaded at http://magma.nationalgeographic.com/ngm/0210/feature1/map.html (accessed August 22, 2004); and information on "The Ibis Galleries" is available at http://www.aldokkan.com/geography/serapeum.htm (accessed August 22, 2004).

74. Here is what Pliny wrote: "THE AEGYPTIANS likewise have recourse in their praiers and invocations to their birds named Ibis, what time as they be troubled and annoied with serpents comming among them. And in like case the Eleans seeke unto their god *Myiagros*, for to be rid of a multitude of flies which pester them so, that they breed a pestilence among them. But looke upon what day they find that Idoll appeased and pacified by their sacrifice, all the flies die forthwith." C. Plinius Secundus, "Of the Bird Ibis," in Pliny, *The Naturall Historie of the World (1601)*, trans. P. Holland, book X (chapter XXVIII), pp. 270–309, available at http://penelope.uchicago.edu/holland/pliny10.html#chap28 (accessed August 26, 2004). See also Pliny, *Natural History*, vol. 3, book VIII (chapter XLI), p. 71: "A somewhat similar display has also been made in the same country of Egypt by the bird called the ibis, which makes use of the cure of its beak to purge itself through the part by which it is most conducive to health for the heavy residue of foodstuffs to be excreted." That is, it relieves itself of the pressure from having overeaten. And see K. Clark, *Animals and Men*, 75; Gosling, Manti, and Nicholson, "Discovery and Conservation of Votive Bronzes"; Ray, "World of North Saqqara"; C. Stetter, *The Secret Medicine of the Pharaohs: Ancient Egyptian Healing* (Chicago: Edition Q, 1993), 32, 35; *Herodotus: The Histories*. trans. A. de Sélincourt (Baltimore, MD: Penguin, 1954), 131.

75. H. A. Klein and M. C. Klein, *Peter Brueghel the Elder* (New York: Macmillan, 1968), 9–10, 12–15.

76. For information on the return of the Osprey to Scotland see http://www.ospreys.org.uk/About%20Scotland.htm (accessed August 26, 2004); Ehrlich, Dobkin, Wheye, and Pimm, *Birdwatcher's Handbook*, 524; F. Grossmann, *Pieter Brueghel: Complete Edition of the Paintings*, 3rd ed. rev. (New York: Phaidon, 1973), 202. Grossman also lists the iconography and proverb portrayed in Plate 34. Quoting the Brueghel scholar G. Hulin de Loo, Grossmann writes: "He who knows where the nest is, has the knowledge, he who robs it has the nest." Another translation by Hulin de Loo, quoted in another book about Brueghel and also appearing in the Gallery Guide in this book (see Plate 34), takes the form of a rhyme: "The one who knows the nest's location, / Can say that he has known. / The one who steals the nest, however, / Has it for his own." The rhyme is followed by a clarification that the walking peasant is the "knower" and the climbing peasant the "possessor." Klein and Klein, *Peter Brueghel the Elder*, 168.

77. H. Bub, *Bird Trapping and Bird Banding: A Hand-*

book for Trapping Methods All Over the World, trans. Frances Hamerstrom and Karin Wuertz-Schafer (Ithaca, NY: Cornell University Press, 1991), caption for the frontispiece: "The woman banding the heron is Christine Charlotte Prinzessin von Hessen-Kassel (1725–1782). She is flanked by Freiherrn Ludolph von und zum Canstein (left) and L. H. von Osterhausen (right). Onlookers include Ernst Ludwig, Erbprinz von Sachsen-Gotha, Constantin Landgraf von Hessen-Rotenburg and Friedrich W.G. von Oynhausen."

78. For the history of bird banding see http://www.pwrc.usgs.gov/BBL/homepage/history.htm (accessed August 24, 2004).
79. A. Sugden and E. Pennisi, "When to Go, Where to Stop," *Science* 313 (2006): 75; Holden, "Inching Toward Movement Ecology," 779–782.
80. The quotation is from "Guarding the Landscape with Paint—How Robert Bateman Masters the Art of Conservation in Canada's Coastal Forests," *International Wildlife* (July–August 2000), available at http://findarticles.com/p/articles/mi_m1170/is_2000_July-August/ai_62768254 (accessed September 1, 2007). Many examples of Bateman's art can be viewed online at http://www.galleryone.com/bateman_m.htm (accessed 2004).
81. Ehrlich, Dobkin, and Wheye, *Birder's Handbook,* 294; Pimm et al., "Human Impacts on Bird Extinctions," 10945.

Mezzanine

1. Snow, *Two Cultures and the Scientific Revolution,* 3–4, 17. In 1959, C. P. Snow proposed that scientists and artists are members of non-exchanging, opposing cultures. He blamed much of the split on a century long flood of scientific information—one that left a waterline so high it effectively blocked communication with anyone outside one's scientific specialty, let alone across the science-art divide.
2. Root-Bernstein, "Correlations." A description of how dual-career scientist-artists could be accomplished in the 1990s is found in M. Wexler, "Making an Art of Science," *Natural Wildlife* (December–January 1994): 23–26.
3. Wheye, "Science, Art, and Science Art."
4. H. Poincaré, *Science and Method,* trans. Francis Maitland (New York: T. Nelson and Sons, 1914): 129; P. Medawar, *Advice to a Young Scientist* (New York: Harper and Row, 1979). Medawar defined science as "a logically connected network of theories that represents our current opinion about what the natural world is like," noting that it is fed by a hypothesis-driven "search to find the truth using common sense" and adding that where imagination and creativity came in was in formulating the hypothesis.
5. M. Planck, *Scientific Autobiography and Other Papers,* trans. Frank Gaynor (1949; reprint, Westport, CT: Greenwood Press, 1968), 109; A. Einstein, *The Evolution of Physics* (New York: Simon and Schuster, 1938), 92.
6. Janson, *History of Art,* 593 (observable facts); Gombrich, *Art and Illusion,* 328 ("unite imagination with nature"); R. B. Beckett, ed. and comp., *John Constable's Discourses* (Ipswich, England: Suffolk Records Society / W. S. Cowell, 1970), 39 ("comparison with realities"). A comment posted on the Web: "Constable painted the world as he saw it—not an idealized, classical landscape painted from on high, but a working rural scene, with a canal, a field and a mill, as seen by the viewer about to walk along the towpath. . . . Constable was not the first artist to paint a landscape from the ground up (Jacob van Ruisdael pioneered this approach in the 1600s), but he was the first artist in England to bring landscape down to earth; and at first he hit the ground with a thud. The prevailing taste held that landscape was inferior to history painting. While J. M. W. Turner invented historical landscapes that were acceptable to the Royal Academy of Arts (RA), then Britain's arbiter of taste and primary exhibition space, Constable's earthy landscapes, based both on his experience and his deeply held religious beliefs in nature as the site of Revelation, were a harder sell." This comment, posted on June 1, 2006, appeared in a discussion of the exhibition "Richard Long and John Constable: Lords of the Landscape" from the *Economist* print edition; an online version is available to subscribers of the *Economist* at http://www.economist.com/cities/PrinterFriendly.cfm?Story_ID=7001020 (accessed June 21, 2006). To view examples of Constable's work see S. Cove, J. Gage, A. Lyles, and C. Rhyme, *Constable: The Great Landscapes* (London: Tate, 2006).
7. *Paul Klee Notebooks,* ed. J. Spiller, trans. Ralph Manheim (New York: G. Wittenborn, 1961–1973).
8. Quoted in A. Koestler, *The Act of Creation: A Study of the Conscious and Unconscious in Science and Art* (New York: Dell, 1964), 148.
9. L. B. Meyer, *Emotion and Meaning in Music* (Chicago: Chicago University Press, 1956), 70; Koestler, *Act of Creation,* 252.
10. Tauber, *Elusive Synthesis,* 9.
11. Holt, *Search for Aesthetic Meaning,* 66.
12. E. H. Gombrich, "Miracle at Chauvet," *New York Review of Books* (November 14, 1996): 10, 12.
13. W. H. McNeill, "Secrets of the Cave Paintings" (review), *New York Review of Books* (October 19, 2006), available online at http://www.nybooks.com/articles/19435.
14. Gombrich, *Art and Illusion,* 330–331. Readers might like to consider two additional points. First, the ancient Egyptians often portrayed people in profile. John Hyman points out that the Egyptian hieroglyph of the human head in profile conveyed the idea of the head, whereas the hieroglyph of the human head presented "head on" conveyed the idea of just the face. He also notes that Egyptian art does not employ shadows. J. Hyman, *The Imitation of Nature* (Oxford: Basil Blackwell, 1989), 106–107. Second, Nancy Aiken has built an argument for a biological origin of art that has something to say about "conventional" lines or forms, beginning with the assumption that art evokes a wide variety of emotional responses, some of which are automatic and univer-

sal. N. E. Aiken, *The Biological Origins of Art* (Westport, CT: Praeger, 1998), 3–6, 174.

Upper Gallery

1. Studies have shown that birds are capable of "sleeping" one cerebral hemisphere at a time, allowing them to keep one eye open as they doze. N. C. Rattenborg, C. J. Amianer, and S. L. Lima, "Behavioral, Neurophysiological and Evolutionary Perspectives on Unihemispheric Sleep," *Neuroscience and Biobehavioral Reviews* 24:8 (2000): 817-42. This study is available to subscribers online at http://www.sciencedirect.com/science?_ob=ArticleURL&_udi=B6T0J-41WJBG4-3&_user=145269&_coverDate=12%2F31%2F2000&_rdoc=1&_fmt=&_orig=search&_sort=d&view=c&_acct=C000012078&_version=1&_urlVersion=0&_userid=145269&md5=b951e8c163380717742357e6ed75216a.
2. S. Lerer, "Saying and Meaning: A Plea for the Humanities at Stanford," *Stanford Report* (October 8, 2003): 4.
3. Shepard, *Traces of an Omnivore*, x.
4. Okri, *Way of Being Free*, 112.
5. Lee, *Rock Art of Easter Island*, 18.
6. Ibid., 16.
7. Ibid., 19.
8. F. X. Clines, "As Their Numbers Soar, Birders Seek Political Influence to Match," *New York Times* (February 4, 2001), available at http://www.nytimes.com/2001/02/04/national/04BIRD.html (accessed September 25, 2006). On the National Audubon Society's assessment of public interest in birds see www.audubon.org/bird/wb/html (accessed 2004). On Easter Island see Diamond, *Collapse*, 79–119. Jared Diamond says that Easter Island, once "home to dozens of species of trees, which created and protected an ecosystem fertile enough to support as many as thirty thousand people," is today "a barren and largely empty outcropping of volcanic rock. What happened? Did a rare plant virus wipe out the island's forest cover? Not at all. The Easter Islanders chopped their trees down, one by one, until they were all gone." See also a review of the book: M. Gladwell, "The Vanishing: In 'Collapse' Jared Diamond Shows How Societies Destroy Themselves," *New Yorker* (January 3, 2005), at http://www.newyorker.com/critics/books/?050103crbo_books (posted December 27, 2004).
9. M. D. Breed, E. Guzmán-Novoa, and G. J. Hunt, "Defensive Behavior of Bees: Organization, Genetics, and Comparisons with Other Bees," *Annual Review of Entomology* 49 (2003): 271–298; J. L. Gould and C. G. Gould, *The Honey Bee* (New York: Science American Library, 1988), 4–5.
10. H. Friedmann, "The Honeyguides," *U.S. National Museum Bulletin 208* (Washington, DC: Government Printing Office, 1955), 1–292.
11. Museum of Modern Art, *MoMA Highlights*, rev. ed. (New York: Museum of Modern Art, 2004), 127. The image and the caption can be viewed online at http://www.moma.org/collection/browse_results.php?object_id=37347 (accessed June 1, 2006).
12. Richardson, *Modern Art and Scientific Thought*, 135.
13. S. Hunter, *The Museum of Modern Art, New York: The History and the Collection* (New York: Harry N. Abrams, 1984), 455.
14. Root-Bernstein, "Sciences and Arts Share Aesthetic," 69–71; K. Lorenz, *King Solomon's Ring* (New York: Crowell, 1952), 12; R. S. Root-Bernstein, "Art Fosters Science: New Paradigm for Stimulating Innovation," abstract for a lecture, January 10, 2003, at Scientific Computing and Imaging Institute, University of Utah, available online at http://www.sci.utah.edu/cgi-bin/ALLseminars.pl?display=2003011000 (accessed September 12, 2006). Leonardo da Vinci's notebook, *Codex Atlanticus* (C.A. 161ra), dated c. 1505, is held in the Biblioteca Ambrosiana in Milan. Several facsimiles are listed in the bibliography for M. Kemp, *Leonardo da Vinci: The Marvelous Works of Nature and Man* (Oxford: Oxford University Press, 2006), 350. The notebook is also available as an electronic resource (CD-ROM) with an accompanying book: M. Taddei and E. Zanon, *Codex Atlanticus = Codice Atlantico / Leonardo da Vinci*, Edition: 2.9 (Milan: Leonardo3 srl, 2005).
15. An English translation of Roger Bacon's description of a flying machine, which appeared in *Epistola de secretis operibus artis et naturae* (1608 and 1618), is available in *Roger Bacon's Letter Concerning the Marvelous Power of Art and of Nature and Concerning the Nullity of Magic*, trans. T. L. Davis (Easton, PA: Chemical Pub. Co., 1923), 27.
16. S. Bramley, *Leonardo: Discovering the Life of Leonardo da Vinci*, trans. D. Reynolds (New York: HarperCollins, 1991), 454 n. 88 (quotations); M. Kemp, "The Flying Machine," in *Leonardo da Vinci: Hayward Gallery, London, 26 January to April 1989* (New Haven: Yale University Press in association with the South Bank Centre, London, 1989), 238–239.
17. *Daedalus 88*, designed by students at MIT to cover the 71-mile (114-kilometer) distance that Daedalus needed to escape from King Minos, flew 74 miles (119 kilometers) from Crete to the island of Santorini in just under four hours in 1988. The aircraft, which crashed in the sea just short of Santorini, weighed 69 pounds (31 kilograms). Additional information and images are available at http://www.dfrc.nasa.gov/Gallery/Photo/Daedalus/HTML/ (accessed September 26, 2006).
18. W. H. Adams, ed., *The Eye of Thomas Jefferson* (Washington, DC: National Gallery of Art, 1976), 61–62; Uglow, *Lunar Men*, xiv, xx, 122; Nicholson, *Joseph Wright of Derby*, 2:114–115. For information on the English city of Derby see http://www.derbyuk.net/wright.html (accessed August 14, 2004).
19. Nicholson, *Joseph Wright of Derby*, 1:111–114, 2:44–45. Guericke used the pump to study partial vacuums and combustion. More information on Guericke and his air pump is available online in such Google Books as H. Smith and E. H. Williams, *A History of Science* (New York: Harper and Brothers, 1904), 211.
20. Nicholson, *Joseph Wright of Derby*, 1:36.
21. A. Boyle, "From Quasicrystals to Kleenex," *Plus:*

(September 2001), available at http://plus.maths.org/issue16/features/penrose/ (accessed February 8, 2005). Alison Boyle explains Penrose tiles in detail in this online publication hosted by Cambridge University. The quotation is from F. H. Bool, J. R. Kist, J. L. Locher, and F. Wierda, *M. C. Escher: His Life and Complete Graphic Work* (New York: Harry N. Abrams, 1981), 100.

22. Jones and Galison, *Picturing Science, Producing Art*, 20, 2–3.
23. M. McLuhan, *Understanding Media: The Extensions of Man* (Cambridge, MA: MIT Press, 1994), 69.
24. Prey selection depends not only on how difficult the prey is to find, catch, and eat but on how much energy it will provide relative to the risks the predator takes during its capture. For active snakes additional factors include the difficulty of slithering on branches of various thicknesses, limitations posed by microclimates, and other hazards associated with specific foraging sites, such as distance from the ground. Inactive snakes (sit-and-wait snakes) generally depend on visual or olfactory cues for locating prey, although some use heat-sensitive pits. A 1983 study of prey taken by tropical arboreal snakes found that 53 percent ate frogs, 65 percent ate reptiles, 43 percent ate birds and eggs, and 26 percent ate mammals. R. A. Seigel and J. T. Collins, *Snakes: Ecology and Behavior* (New York: McGraw-Hill, 1993), 24–33.
25. Tauber, *Elusive Synthesis*, 7.
26. T. Angell, pers. comm., December 23, 2004.
27. G. Orians, *Blackbirds of North America* (Seattle: University of Washington Press, 1985).
28. For information on the role of diagrams see Freedberg, *Eye of the Lynx*, 97. By the beginning of the seventeenth century, natural history collections were expanding at an accelerated rate, and warehousing representative specimens of nature's diversity required space. The development of "paper museums" illustrating specimens gained favor, and scientific illustration, which had been increasingly in demand in response to improvements in navigation and seagoing explorations, gained momentum. Much credit is owed to Federico Cesi (1585–1630), a devotee of Galileo, and to Cesi's Academy. A. S. Blum, *Picturing Nature: American Nineteenth-Century Zoological Illustration* (Princeton, NJ: Princeton University Press, 1993), 3–4. Blum cites the use of schemata and diagrams as an alternative to realism in the nineteenth century. The expansion of museum collections and the increase in laboratory studies—as opposed to field studies—as well as changes in printing technology, led to new conventions. See also the coverage of Cesi in D. Attenborough, S. Owens, M. Clayton, and R. Alexandratos, *Amazing Rare Things: The Art of Natural History in the Age of Discovery* (New Haven: Yale University Press, 2005). For information on Norman Rockwell see http://www.illustration-house.com/bios/rockwell_bio.html (accessed September 14, 2004).
29. E. Scott, "Mr. and Mrs. Norman Rockwell: Shaker Festival Exhibitors," *Albany (NY) Times Union* (July 27, 1969).
30. The discussion here is adapted from material written by Linda Pero, curator of Norman Rockwell Collections, for a small "case" exhibit; and from "Art with Accuracy," in "Living with People," *McCall's* (April 1967). For information on Mozart's starling see http://www.starlingtalk.com/mozart1.htm (accessed 1–11–05). For more on the Presidential Medal of Freedom see http://www.presidency.ucsb.edu/ws/index.php?pid=5550.
31. The study by E. R. Kalmbach is reported in the sixty-four-page *Birds in Relation to the Alfalfa Weevil*, U.S. Department of Agriculture Bulletin 107 (Professional Paper) (July 27, 1914), and is discussed in *The Condor* (March 1915): 108.
32. F. G. Marcham, "The Hands of an Angel," *Living Bird* (Autumn 1989): 11–13; Peck, *Celebration of Birds*, 22. The authors checked various cards for Fuertes' signature, including one published in Marcham's article and others on the Internet, and did find his signature. Fuertes' watercolors in *The Bird Life of Texas*, which includes some of the same paintings, show his signature as well: H. C. Oberholser with paintings by Louis Agassiz Fuertes, *The Bird Life of Texas* (Austin: University of Texas Press, 1974). Fuertes' recognition is noted in R. T. Peterson, *The Bird Watcher's Anthology* (New York: Bonanza, 1957), 267.
33. Oberholser, *Bird Life of Texas*, 16; M. F. Boynton, ed., *Louis Agassiz Fuertes: His Life Briefly Told by His Correspondence* (New York: Oxford University Press, 1956), 212.
34. Robert Bateman went on to say, "Real art was about itself, whereas illustration is about something else. To me, illustration has a clear meaning. It is when the idea comes from outside the artist, as an assignment." The quotation is available at http://www.batemanideas.com/art.html (accessed September 26, 2006).
35. A brief biography of Roger Tory Peterson is available at http://www.rtpi.org/?page_id=21 (accessed September 2, 2007).
36. M. S. Livingston, "Is It Warm? Is It Real? Or Just Low Spatial Frequency?" *Science* 290 (2000): 1299. Wearing a gauzy overdress was traditional among sixteenth-century Italian women who were nursing. Its presence in the painting helps identify the subject as the wife of a silk merchant who commissioned the work to commemorate the birth of their second child in 1502. It also helps resolve speculation about self-portraiture (the slight resemblance), tooth decay (the tight smile), and stroke (the passive arm). I. Austen, "New Look at 'Mona Lisa' Yields Some New Secrets," *New York Times* (September 27, 2006), available at http://www.nytimes.com/2006/09/27/arts/design/27mona.html?ref=arts (accessed September 27, 2006); Livingston and Conway, "Was Rembrandt Stereoblind?"
37. J. A. Ferwerda, "Three Varieties of Realism in Computer Graphics," in B. E. Rogowitz and T. N. Pappas, eds., *Proceedings SPIE: Human Vision and Electronic Imaging VIII* (International Society for Optical Engineering, 2003), 290–297; the article also is available at www.graphics.cornell.edu/~jaf/publications/vor_hvei03_v20.pdf (accessed September 2, 2007). Fer-

werda notes on the first page that "making an image involves processes of selection, approximation, and abstraction."

38. Rhodes, *John James Audubon*, 111.
39. For the text of the film see J. Perrin, *Winged Migration (Le Peuple Migrateur)*, text by J-F. Mongibeaux, trans. D. Wharry (San Francisco: Chronicle; London: Hi Marketing, 2003), 248. The film was nominated for an Oscar in 2003.
40. Perrin, *Winged Migration*, 242.
41. H. Bismuth, quoted in *Birds in Art: 1995* (Wausau, WI: Leigh Yawkey Woodson Art Museum, 1995), 31. Corvids are member of the Corvidae family, a group of highly intelligent, harsh-voiced, often gregarious birds with all-purpose bills. The family also includes crows, jays, and magpies.
42. B. Heinrich, *The Mind of the Raven*, 191–205; T. Angell, *Ravens, Crows, Magpies, and Jays* (Vancouver, Canada: Douglas and McIntyre, 1978), 84.
43. M. G. Zackowitz, "Wildlife," *National Geographic* (April 2006).
44. P. Waterman, quoted in *Birds in Art: 1997* (Wausau, WI: Leigh Yawkey Woodson Art Museum, 1997), 116.
45. Ehrlich, Dobkin, and Wheye, *Birder's Handbook*, 420. For more on the raven see http://www.birdweb.org/birdweb/species.asp?id=318 (accessed September 15, 2004).
46. C. Brenders, quoted in *Birds in Art: 1998* (Wausau, WI: Leigh Yawkey Woodson Art Museum, 1998), 30.
47. The lithograph, one of Roger Tory Peterson's early ones, was commissioned by the Atlantic Company, a manufacturer of picture frames that had taken over the publishing of prints from the Quaker State Lithographing Company. None of Peterson's correspondence with these companies mentions the grosbeaks, and the whereabouts of the original is unknown. For more on Roger Tory Peterson and the Evening Grosbeak see http://birdcare.com/bin/shownews/58 (accessed August 28, 2004).
48. For online information on the Evening Grosbeak see http://www.mbr-pwrc.usgs.gov/id/framlst/button/lh5140.html (accessed January 27, 2005); and http://birds.cornell.edu/BOW/evgr/ (accessed January 27, 2005). See also F. M. Chapman, *Birds of Eastern North America* (New York: D. Appleton, 1932), 510; R. T. Peterson, *A Field Guide to the Birds* (Boston: Houghton Mifflin, 1959), 223; J. Bull and J. Farrand, Jr., *The Audubon Society Guide to North American Birds* (New York: Knopf, 1979), 694.
49. For online information on the Evening Grosbeak see http://birds.cornell.edu/BOW/evgr/ (accessed January 27, 2005).
50. Information on Roger Tory Peterson is included in the National Audubon Society's *Annual Report for 2005*, at ct.audubon.org/PDFs/Audubon05.pdf (accessed September 10, 2007). In a 2007 study, released on June 14, the National Audubon Society reported that the Evening Grosbeak has declined by 78 percent since 1967. Among common birds, this is the second largest decline in North America. According to the author, Greg Butcher, besides global climate change, causes include suburban sprawl and invasive species.
51. The first two quotations are from Patricia Pépin, pers. comm., January 12, 2005; the third is from Pépin, quoted in *Birds in Art: 1998*, 81.
52. The first, third, and fourth quotations are from Patricia Pépin, pers. comm., January 12, 2005; the second is from Pépin, quoted in *Birds in Art: 1998*, 81.
53. This quotation is also from Patricia Pépin, pers. comm., January 12, 2005.
54. The two quotations are from material received from Carolyn Ching, pers. comm., December 30, 2004. They are from published accounts about this painting: R. H. Ching, *Voices from the Wilderness: Paintings and Drawings* (Shrewsbury, UK: Swan Hill, 1994), 174–179.
55. J. Fisher, "Is There a Message in These Paintings?" *National Wildlife* (June–July 1995): 50. The quotation is also in material received from Carolyn Ching, December 30, 2004.
56. A brief biography of Ray Harris Ching is available online at http://www.wildlifeart.org/Collections/ArtistBio.cfm?tArtistid=311&tUrl=ArtistBioMain.cfm&UrlName=Artist%20Biographies (accessed September 20, 2004).
57. T. Quinn, quoted in *Birds in Art: 1997*, 84. The other quotations are from Rayfield, *Wildlife Painting Techniques of Modern Masters*, 197.
58. The quotations in this caption are from Tom Grey, pers comm., December 24, 2004.
59. P. Hayman, J. Marchant, and T. Prater, *Shorebirds: An Identification Guide to the Waders of the World* (Boston: Houghton Mifflin, 1986), 321.
60. C. Bacon, quoted in *Birds in Art: 1999* (Wausau, WI: Leigh Yawkey Woodson Art Museum, 1999), 25. Bacon's other quotations in the caption are from his Web site, www.chrisbacon.com.
61. Ehrlich, Dobkin, and Wheye, *Birder's Handbook*, 125, 127.
62. Artist Registry for Ornithological Researchers, available at http://artist-registry.stanford.edu (accessed September 13, 2006). All of Harper's quotations in the caption are from www.fabframes.com.
63. P. R. Ehrlich, D. S. Dobkin, and D. Wheye, *Birds in Jeopardy: The Imperiled and Extinct Birds of the United States and Canada, Including Hawaii and Puerto Rico* (Stanford, CA: Stanford University Press, 1992), 22.
64. D. Thompson and I. Byrkjedal, *Shorebirds* (Stillwater, MN: Voyager Press, 2001), 55. According to Elliott Coues, ornithologist and cofounder of the American Ornithological Union (AOU), the curlews ate berries until their "intestines, the vent, the legs, the bill, throat, and even the plumage [were] more or less stained with the deep purple juice." E. Coues, "Notes on the Ornithology of Labrador," *Proceedings of the Academy of Natural Sciences of Philadelphia* 13 (1861): 237.
65. Holt, *Search for Aesthetic Meaning*, 71.
66. Serrell, *Exhibit Labels*, 233–234.
67. Ibid., 234–236.
68. Ehrlich, Dobkin, and Wheye, *Birder's Handbook*, 312.

69. A brief bibliography featuring Brest van Kempen's work appears in Appendix 2.
70. Peterson and Peterson, *Audubon Society Baby Elephant Folio,* plate 307. Roger Tory Peterson notes in the caption, "Critics correctly point out that rattlesnakes are not tree climbers, and the birds are more likely to nest in shrubs than in trees." He continues, "Mockingbirds are, however, well-known for the wing-flashing clearly shown here, a behavior that, ostensibly, serves to distract predators, especially snakes." Point taken, but the story still has legs. The American International Rattlesnake Museum in Albuquerque, New Mexico, for example, writes, "Drama and design come together in one of his most famous, and controversial, plates—mockingbirds defending a nest against a timber rattler. Claiming that rattlesnakes do not climb trees, and that mockingbirds do not come to the aid of another nest, Audubon's enemies attacked him. As it turns out, rattlesnakes can climb, but rarely do. This is one of only 4 snakes that Audubon painted in his lifetime." www.rattlesnakes.com (accessed December 23, 2004). Audubon also praised the vocal mimicry of this bird, writing, "There is probably no bird in the world that possesses all the musical qualifications of this king of song." Many would nominate members of the starling and thrasher families.
71. L. Kostrich, pers. comm.; www.minniesland.com (accessed September 26, 2006).
72. Ford, *Images of Science,* 56; T. H. Clarke, *The Rhinoceros: From Dürer to Stubbs—1515–1799* (London: Sotheby's, 1986), 20.
73. F. Koreny, *Albrecht Dürer and the Animal and Plant Studies of the Renaissance* (Boston: New York Graphic Society, 1986), 84–92. Many people would refer to the Hoffman image as a watercolor, but museums often classify watercolors as drawings because they are painted on paper. Referring to the Hoffman "drawing" on exhibit, the National Gallery's curator of old master drawings Margaret Morgan Grasselli noted, "You can see a difference if you put it side by side with Dürer's original in Vienna, but it's a great picture itself." C. Hartman, "National Gallery Opens Masters Exhibit" (May 1, 2006), at http://www.artinfo.com/articles/story/15331/national_gallery_opens_old_masters_exhibition (accessed September 12, 2007).
74. Olaf Worm's Copenhagen Museum was one of the first to include birds. Bird specimens were difficult to warehouse because preserving methods were primitive and specimens were quickly lost to infestations and decay, a situation that would not significantly improve for generations. On the Great Auk, Fuller, in *Great Auk,* 361, writes: "A portrait of his bird was drawn and this much reproduced illustration is one of the earliest known representations of the species. A peculiarity is a narrow, white band around the throat—perhaps an embellishment of the artist, perhaps a collar by which the captive was led around." See also J. Kastner and M. Gross, *The Bird Illustrated, 1550–1900: From the Collections of the New York Public Library* (New York: Harry N. Abrams, 1988); Kastner, "The Bird Illustrated: With the Renaissance Came Science and Printing—and an Explosion of Ornithological Art," *Natural History* (March 1988): 33–38.
75. For more on Stanford Art Spaces see http://cis.stanford.edu/~marigros/ (accessed October 16, 2004). Over the past two decades the Stanford Art Spaces have accommodated more than 100 exhibits representing more than 200 artists and more than 5,000 works of art. The program offers numerous advantages. The university installs and insures the artwork, publicizes the exhibits, and allows sales. The selection process resembles that for most galleries, where interested participants submit a resume along with sample slides or photographs.
76. If it were possible to eliminate "science, art" from an Internet search, the results would be considerably smaller. Instead, a search lists entries that relate to science or to art, but not to Science Art and not to the artwork itself. A Google search on "'Science Art' Nature" yielded 1,040,000 results in April 2007, up from 43,300 on October 25, 2004; and one on "Science Art Nature" yielded 494 results on September 25, 2006, up from 318 on October 25, 2004. Even those results, however, are unlikely to include Web sites featuring examples of Science Art, and the few that are included are unlikely to be found easily. For people interested in locating Science Art online, the authors have posted a Web page, "Finding Science Art Online," accessible via www.scienceart-nature.org, which includes suggestions and links. Additional comments are provided in Appendix 2.
77. Proffitt and Bagla, "Circling In on a Vulture Killer," 223. Two years later, in May 2006, it was reported that Pakistan did not switch from diclofenac to meloxicam, an alternative treatment that does not harm Oriental White-backed Vultures, and lost three of its five breeding colonies, with the number of breeding pairs dropping from 3,500 to 75. R. Koenig, "Vulture Research Soars as the Scavengers' Numbers Decline," *Science* 312 (June 2006): 1591–1592.
78. K. Brower, "Night of the Condor," first published in *Omni* (1979), reprinted in *Not Man Apart* (February 1980). A report by John Nielsen on NPR's *All Things Considered* (June 26, 2002) is available at http://www.npr.org/programs/atc/features/2002/june/vultures/ (accessed November 6, 2004). See also the National Geographic's report at http://news.nationalgeographic.com/news/2004/01/0128_040128_indiavultures.html (accessed November 6, 2004).
79. For a discussion of realism in seventeenth-century Dutch painting see Freedberg and de Vries, *Art in History, History in Art,* 2.
80. Channing Robertson, pers. comm.
81. The production of a database could be greatly advanced with the involvement of scientific organizations like the American Association for the Advancement of Science (AAAS), the world's largest general scientific society and the publisher of *Science* magazine, and the American Institute of Biological Sciences (AIBS), which includes seventy-eight member societies and organizations. The Council of Science Editors Web site provides links

at http://www.councilscienceeditors.org/services/societylinks.cfm (accessed November 1, 2004).

82. Science Art is not recognized as a subject heading by the Library of Congress, the Society of Animal Artists, or the Watercolor Society, nor in the Getty *Art and Architecture Thesaurus* or *The Grove Dictionary of Art,* nor in the online databases ARTbibliographies Modern, ARTstor, or the Art Museum Image Consortium (AMICO). The authors are contacting a long list of competitions, exhibitions, galleries, museums, art magazines, and nature and bird organizations both in the United States and abroad to explain the value of making it a category. For a list of galleries that regularly exhibit bird art, see The Artists Registry for Ornithological Researchers, which is accessible at birds.stanford.edu.

83. Parents of young children might draw on books that have stood the test of time, like the 100 titles by Thornton Burgess (1910–1965), as well as a wealth of new multimedia resources and books that come with seals of approval in the form of annual awards. An example is *Daniel and His Walking Stick,* written by Wendy McCormick and illustrated by Constance R. Bergum, which won the 2006 Giverny Award for the best children's science picture book; it shows how one generation effectively passes along knowledge about nature to the next. The best books for students in K–12 also come with seals of approval. They are occasionally written by authors who ordinarily write for adults, as is the case with Carl Hiaasen's *Hoot*—a middle-school mystery about Burrowing Owls and eco-avengers, which became a 2003 Newbery Medal Honor Book and a 2006 motion picture. Collaboration between authors and artists producing nature-related Science Art for children can effectively introduce them to elements of the environment, but an international online registry of artists specializing in this kind of art for children has yet to be established.

84. The Environmental Literacy Council received a three-year grant from the National Science Foundation for a joint project with the National Science Teachers Association to produce resources for middle- and high-school science teachers that illustrate the application of core science ideas to environmental issues. The council also received a two-year grant from the National Endowment for the Humanities to produce a series of environmental history teaching guides in collaboration with the National Humanities Center. "Our classrooms must become places where students achieve a deep understanding of complex environmental issues. A forest, for example, may be at one and the same time a place of great beauty; a natural resource critical to the health and well-being of neighboring communities; a local ecosystem, supporting rich plant and animal life; and a vital component in the planet's great biogeochemical cycles for regulating global climate. The Council seeks to help teachers and their students see this forest and its trees: to analyze and evaluate risk, and to understand the limits and impact of our actions." The council's Web site is http://www.enviroliteracy.org/index.php (accessed October 30, 2004). Teachers of upper-level environmental science courses can share exam questions with colleagues and use the Environmental Science Testbank to create online environmental science tests for students, selecting questions by topic and form (true/false, multiple choice, short answer, or essay) from a list of submitted questions.

85. Information on the emergence of science in the modern sense of the term is readily available through online and print encyclopedias. Penny Olsen's coverage of Australian bird art in *Feather and Brush* (2001) includes the work of 100 artists and highlights similarities between Australian and U.S. bird art. For example, both Neville W. Cayley's and Roger Tory Peterson's field guides, published in 1931 and 1934 respectively, marked the beginning of modern birding in those countries, and the lull in the production of bird art between World War II and the rise of the Environmental Movement in the 1960s (which some credit to the expanded popularity of nature photography) gave way to a renaissance of sorts with the widespread exhibition of images depicting jeopardized species. Other well-documented books featuring bird art include those by Christine Jackson, Roger Pasquier and John Farrand, Jr., and A. M. Lysaght. See also Appendix 1 and Selected Bibliography.

86. Hoose, *Race to Save the Lord God Bird,* 26.

87. D. R. Eckelberry, "Search for the Rare Ivorybill," in J. K. Terres, ed., *Discovery: Great Moments in the Lives of Outstanding Naturalists* (Philadelphia: J. B. Lippincott, 1961), reprinted as "On the Heels of the Dodo," in H. Borland, ed., *Our Natural World: The Land and Wildlife of America as Seen and Described by Writers since the Country's Discovery* (Philadelphia: Lippincott, 1961), 642–643.

88. Rhodes, *John James Audubon,* 13. For a work of art showing actual supports, see Sargent's *Studies of a Dead Bird* (1878), which is included in D. B. Burke, *American Paintings in the Metropolitan Museum of Art: A Catalogue of Works by Artists Born between 1846 and 1864,* vol. 3, ed. K. Luhrs (New York: The Metropolitan Museum of Art; Princeton, NJ: Princeton University Press, 1980), 222.

89. A. Wilson and C. L. Bonaparte, *American Ornithology, or The Natural History of the Birds of the United States* (Philadelphia: Bradford and Inskeep, 1808–1814), vol. 1, p. 157.

90. Eckelberry, "Search for the Rare Ivorybill," 644.

91. E. Stokstad, "Gambling on a Ghost Bird," *Science* 317 (2007): 888–892. See the article for a discussion on the odds of the bird's survival.

92. W. Faulkner, *Go Down, Moses* (1942; reprint, New York: Vintage Books, 1973), 202.

93. Janson, *History of Art,* 620–621.

94. Sutton is quoted in Hoose, *The Race to Save the Lord God Bird,* 74. Albert R. Brand, an associate in ornithology at the American Museum of Natural History, wrote an interesting account of the trip to record the Ivory-billed Woodpecker: A. R. Brand, "Bird Voices in the Southland: Making 'Talkies' with an All Star Cast of Native American Birds," *Natural History* (February 1936), available online

at http://www.naturalhistorymag.com/editors_pick/1936_02_pick.html (last accessed July 23, 2006).

95. J. A. Jackson, *In Search of the Ivory-billed Woodpecker* (Washington, DC: Smithsonian Books, 2004), 239, 248.

96. A Google search on "'Ivory-billed Woodpecker' and painting" returned 44,600 results on September 25, 2006, up from 1,330 results on January 25, 2005. The rise in number between 2005 and 2006 indicates that without improvements in narrowing search criteria, image searches, especially for highly jeopardized, iconic species, will become increasingly cumbersome.

Appendix 1. Timeline Linking Art, Technology, and the Study of Birds

1. Rick Bonney coined the term "citizen scientist" in the 1990s.
2. Numerous sources were of particular help in creating the timeline. On art and culture the following should be mentioned: Chilvers, *Concise Oxford Dictionary of Art and Artists*; Elphick, *Birds: The Art of Ornithology*; Ford, *Images of Science*; Hammond, *Modern Wildlife Painting*; Jackson, *Dictionary of Bird Artists of the World*; Janson, *History of Art*; Lysaght, *Book of Birds*; Olsen, *Feather and Brush*; Pasquier, and Farrand, *Masterpieces of Bird Art*; Pollard, *Birds in Greek Life and Myth*; Sitwell, Buchanan, and Fisher, *Fine Bird Books, 1700–1900*; Toynbee, *Animals in Roman Life and Art*. On science the following should be mentioned: Coulson, "Ornithology and Ornithologists in the Twentieth Century"; Mayr, "Materials for a History of American Ornithology"; Stresemann, *Ornithology: Aristotle to the Present*; Walters, *Concise History of Ornithology*.
3. During the whole time the study of birds developed as a science and birding emerged as a popular avocation for hundreds of millions nonprofessionals, artists painted portraits of major innovators, some of whom were artists themselves. Examples of artist-scientist polymaths include Frederick II (1194–1250), John James Audubon (1785–1851), Alexander Wilson (1766–1813), Thomas Bewick (1753–1828), John Gould (1804–1881) and Roger Tory Peterson (1907–1996).
4. Uglow, *Lunar Men*, xx.
5. Norelli, *American Wildlife Painting*, 181. Thayer, even with the assistance of John Singer Sargent in London, had difficulty convincing military leaders to adopt camouflage techniques during World War I. He has been described as particularly upset that "the Germans had developed camouflage technique using [his] theories, before the Allies" would adopt them. By World War II, reluctance was gone.
6. C. Holden, "Eye in the Sky," *Science* 33 (2006): 780–781.
7. For additional information, see, for example, Ehrlich, Dobkin, Wheye, and Pimm, *Birdwatcher's Handbook*, 520–530; Ehrlich, Dobkin, and Wheye, *Birder's Handbook*, 363.

Appendix 2. A Science Art Checklist for Practitioners

1. When the United States joined the Berne Convention for the Protection of Literary and Artistic Works in 1989, multinational copyright protection became automatic, so inclusion of the notice is optional. Thus, adding the words "Science Art" or, more specifically, "Science Art—Nature" or, even more specifically, "Science Art—Birds" to the copyright line should raise no legal red flags.
2. To see where to insert these words in the metatag of the Web page you are viewing, go to your browser's menu bar, pull down the View tab, and click on Source. A window showing the code [Source] will appear. At the top you should find categories with information inserted after them: <html><head><title> </title><meta http-equiv> <meta name="description" content=" "> <meta name="keywords" content="[A]"><LINK REL></head>. Using the proper software (Dreamweaver, BBedit, etc.), insert the keywords "Science Art Nature" at [A].
3. In 2008 the authors began contacting many agencies, requesting that the category be considered for possible future listing. We also began contacting a core selection of art magazines, including *Art Forum*, *Art News*, *Art in America*, *Artweek*, *Art Bulletin*, *Art Newspaper*, *American Artist*, *Communication Art*, *Print*, and *Graphics*, that are well positioned to bring attention to Science Art. In addition, we began contacting the nature and bird organizations sponsoring the Artist Registry for Ornithological Researchers, which are also well positioned to consider including the category in future art competitions and exhibits.
4. When the Artist Registry was launched, the authors and Paul Ehrlich contacted 800 researchers who had published bird-related research, 100 journal editors, 100 galleries and museums featuring art that portrays nature, and international artists specializing in birds and gained nonfunding endorsement from 16 sponsors, including major ornithological organizations, natural history and art museums, academic centers, and major birding and wildlife artist organizations. The Artist Registry features the work of more than 100 bird artists.
5. For North American coverage see, for example, "All About Birds," at the Cornell University Web site http://www.birds.cornell.edu/AllAboutBirds/BirdGuide/ (accessed October 5, 2006); and for global coverage see Avibase, an extensive database in several languages that currently contains more than 2 million records for about 10,000 species and 22,000 subspecies of birds. http://www.bsc-eoc.org/avibase/avibase.jsp?pg=home&lang=EN&id=undefined&ts=undefined (accessed October 5, 2006). For information on conservation status see, for example, the Threatened and Endangered Species System (TESS) of the U.S. Fish and Wildlife Service (USFWS) at http://ecos.fws.gov/tess_public/StartTESS (accessed October 5, 2006); and the IUCN Red List at http://www.iucnredlist.org/ (accessed

October 5, 2006). Art timelines include http://www.metmuseum.org/toah/intro/atr/05sm.htm (accessed October 5, 2006); and http://www.xs4all.nl/~knops/timetab.html (accessed October 5, 2006), which covers from the Sumerians to 1514. Ornithological timelines include that of the Smithsonian Institution, which covers 1846 to the present, available at http://sio.si.edu/History/timeline.cfm (accessed October 5, 2006); that of Cornell University, available at http://rmc.library.cornell.edu/ornithology/ (accessed October 5, 2006); and several on the Hexapedia Web site, including one on ornithology at http://www.hexafind.com/encyclopedia/Ornithology (accessed October 5, 2006); one on influential ornithologists at http://www.hexafind.com/encyclopedia/ List_of_famous_ornithologists (accessed October 5, 2006); one on an ornithological timeline at http://www.hexafind.com/encyclopedia/Timeline_of_ornithology (accessed October 5, 2006); and one on notable biologists at http://www.hexafind.com/encyclopedia/List_of_biologists (accessed October 5, 2006).

Selected Bibliography

Brest van Kempen, C. P. *Rigor Vitae: Life Unyielding; The Art of Carel Pieter Brest van Kempen.* Eagle Mountain, UT: Eagle Mountain Publishing, 2006.

Chilvers, I., ed. *The Concise Oxford Dictionary of Art and Artists.* New York: Oxford University Press, 2003.

Clark, K. *Animals and Men: Their Relationship as Reflected in Western Art from Prehistory to the Present Day.* New York: William Morrow, 1977.

Clark, W. B. *The Medieval Book of Birds: Hugh of Fouilloy's Aviarium.* Binghamton, NY: Medieval and Renaissance Texts and Studies, 1992.

Clayton, P. A. *Chronicle of the Pharaohs.* New York: Thames and Hudson, 1994.

Clottes, J., and J. Courtin. "Neptune's Ice Age Gallery." *Natural History* (April 1993): 64–70.

Coulson, J. "Ornithology and Ornithologists in the Twentieth Century." In M. Walters, *A Concise History of Ornithology.* London: Christopher Helm, 2003.

Diamond, J. *Collapse: How Societies Choose to Fail or Succeed.* New York: Viking, 2005.

Edwards, D. *ArtScience: Creativity in the Post-Google Generation.* Cambridge, MA: Harvard University Press, 2008.

Ehrlich, P. R., D. S. Dobkin, and D. Wheye. *The Birder's Handbook,* New York: Simon and Schuster, 1988.

Ehrlich, P. R., D. S. Dobkin, D. Wheye, and S. L. Pimm. *The Birdwatcher's Handbook.* Oxford: Oxford University Press, 1994.

Elphick, J. *Birds: The Art of Ornithology.* London: Scriptum Editions, 2004.

Feduccia, A. *The Origin and Evolution of Birds.* New Haven: Yale University Press, 1996.

Fischman, J. "Painted Puzzles Line the Walls of an Ancient Cave." *Science* 267 (1995): 614.

Fisher, J. "Is There a Message in These Paintings?" *National Wildlife* (June–July 1995): 44–50.

Ford, B. J. *Images of Science: A History of Scientific Illustration.* New York: Oxford University Press, 1993.

Frederick II. *The Art of Falconry, Being the De Arte Venandi cum Avibus of Frederick II of Hohenstaufen.* Translated and edited by C. A. Wood and F. M. Fyfe. Stanford, CA: Stanford University Press, 1961.

Freedberg, D. *The Eye of the Lynx: Galileo, His Friends, and the Beginnings of Modern Natural History.* Chicago: University of Chicago Press, 2002.

Freedberg, D., and J. de Vries, eds. *Art in History, History in Art: Studies in Seventeenth Century Dutch Culture.* Santa Monica, CA: Getty Center for the History of Art and the Humanities, 1991.

Fuller, E. *Extinct Birds.* Oxford, UK: Oxford University Press, 2000.

———. *The Great Auk.* New York: Harry N. Abrams, 1999.

Gerdts, W. H., and D. J. Wagner. *Natural Habitat: Contemporary Wildlife Artists of North America.* New York: Spanierman Gallery, 1998.

Gombrich, E. H. *Art and Illusion: A Study in the Psychology of Pictorial Representation.* London: Phaidon Press, 1959.

Gould, S. J. "A Lesson from the Old Masters." *Natural History* (August 1996): 16–22, 58–59.

Guthrie, R. D. *The Nature of Paleolithic Art.* Chicago: University of Chicago Press, 2006.

Hammond, N. *Modern Wildlife Painting.* Mountfield, Sussex, UK: Pica Press; New Haven: Yale University Press, 1998.

Heinrich, B. *Mind of the Raven: Investigations and Adventures with Wolf-Birds*. New York: Cliff Street Books, 1999.

Holt, D. K. *The Search for Aesthetic Meaning in the Visual Arts*. Westport, CT: Bergin and Garvey, 2001.

Hoose, P. M. *The Race to Save the Lord God Bird*. New York: Farrar, Straus and Giroux, 2004.

Houlihan, P. F., with S. M. Goodman. *The Birds of Ancient Egypt*. Warminster, England: Aris and Phillips, 1986.

Ingersoll, E. *Birds in Legend, Fable and Folklore*. New York: Longmans, Green, 1923.

Jackson, C. E. *Dictionary of Bird Artists of the World*. Woodbridge, Suffolk, UK: Antique Collectors' Club, 1999.

Janson, H. W. *History of Art*. 3rd ed., revised and expanded by A. F. Janson. New York: Harry N. Abrams, 1986.

Jones, C., and P. Galison, eds. *Picturing Science, Producing Art*. With A. Slaton. New York: Routledge, 1998.

Klingender, F. D. *Animals in Art and Thought to the End of the Middle Ages*. Cambridge, MA: MIT Press, 1971.

Lee, G. *Rock Art of Easter Island: Symbols of Power, Prayers to the Gods*. Los Angeles: UCLA Institute of Archeology, 1992.

Lehrer, J. *Proust Was a Neuroscientist*. Boston: Houghton Mifflin, 2007.

Livingston, M. S., and B. R. Conway. "Was Rembrandt Stereoblind?" *New England Journal of Medicine* 351 (2004): 1264–1265.

Lysaght, A. M. *The Book of Birds: Five Centuries of Bird Illustration*. London: Phaidon, 1975.

Marzluff, J. M., and T. Angell. *In the Company of Crows and Ravens*. New Haven: Yale University Press, 2005.

Mayr, E. "Materials for a History of American Ornithology." In E. Stresemann, *Ornithology: Aristotle to the Present*. Cambridge: Harvard University Press, 1975.

Meyer, L. B. *Emotion and Meaning in Music*. Chicago: University of Chicago Press, 1956.

Morell, V. "Arduous Journeys." *Science* 313 (2006): 783–784.

———. "Stone Age Menagerie." *Audubon* (May–June 1995): 54–62.

Nicholson, B. *Joseph Wright of Derby: Painter of Light*. Vol. 1 (text and catalogue); vol. 2 (plates). New York: Pantheon, 1968.

Norelli, M. R. *American Wildlife Painting*. New York: Watson-Guptill in association with the National Collection of Fine Arts, Smithsonian Institution, Washington, DC, 1974.

Okri, B. *A Way of Being Free*. London: Phoenix House, 1974.

Olsen, P. *Feather and Brush: Three Centuries of Australian Bird Art*. Collingwood, Victoria, Australia: CSIRO, 2001.

Pasquier, R. F., and J. Farrand, Jr. *Masterpieces of Bird Art: 700 Years of Ornithological Illustration*. New York: Abbeville, 1991.

Peck, R. M. *A Celebration of Birds: The Life and Art of Louis Agassiz Fuertes*. New York: Walker, 1982.

Peterson, R. T., and V. M. Peterson. *The Audubon Society Baby Elephant Folio: Audubon's Birds of America*. New York: Abbeville, 1981.

Pimm, S., P. Raven, A. Peterson, Ç. H. Şekercioğlu, and P. R. Ehrlich. "Human Impacts on the Rates of Recent, Present, and Future Bird Extinctions." *Proceedings of the National Academy of Science* 103.29 (2006): 10941–10946.

Pliny the Elder. *Natural History, with an English Translation, in Ten Volumes*. Translated by H. Rackham. Vol. 3, Books VIII–XI. Loeb Classical Library. Cambridge: Harvard University Press, 1938.

Pollard, J. *Birds in Greek Life and Myth*. London: Thames and Hudson, 1977.

Proffitt, F., and P. Bagla. "Circling In on a Vulture Killer." *Science* 306 (2004): 223.

Rayfield, S. *Wildlife Painting Techniques of Modern Masters: Famous Artists Show What and How They Paint*. New York: Watson-Gulpill, 1985.

Read, H. *Icon and Idea: The Function of Art in the Development of Human Consciousness*. New York: Schocken, 1965.

Rhodes, R. *John James Audubon: The Making of an American*. New York: Knopf, 2004.

Richardson, J. A. *Modern Art and Scientific Thought.* Urbana: University of Illinois Press, 1971.

Root-Bernstein, R. S. "Correlations between Avocations, Scientific Style and Professional Impact of Thirty-eight Scientists of the Eiduson Study." *Creativity Research Journal* 8 (1995): 115–137.

———. "The Sciences and Arts Share a Common Creative Aesthetic." In A. I. Tauber, ed., *The Elusive Synthesis: Aesthetics and Science.* Boston: Kluwer Academic, 1996.

Ruspoli, M. *The Cave of Lascaux: The Final Photographs.* New York: Harry N. Abrams, 1987.

Said, R. *The River Nile.* Oxford, UK: Pergamon, 1993.

Serrell, B. *Exhibit Labels: An Interpretive Approach.* Walnut Creek, CA: AltaMira Press, 1996.

Shepard, P. *Traces of an Omnivore.* Washington, DC: Island Press, 1996.

Sitwell, S., H. Buchanan, and J. Fisher. *Fine Bird Books, 1700–1900.* New York: Atlantic Monthly Press, 1990.

Snow, C. P. *The Two Cultures and the Scientific Revolution.* The Rede Lecture. New York: Cambridge University Press, 1959.

Stafford, B. M. *Artful Science: Enlightenment, Entertainment, and the Eclipse of Visual Education.* Cambridge, MA: MIT Press, 1994.

Stresemann, E. *Ornithology: Aristotle to the Present.* Translated by H. J. Epstein and C. Epstein. Cambridge: Harvard University Press, 1975.

Strosberg, E. *Art and Science.* New York: Abbeville, 2001.

Tauber, A. I., ed. *The Elusive Synthesis: Aesthetics and Science.* Boston: Kluwer Academic, 1996.

Thomas, C. D., A. Cameron, R. E. Green, M. Bakkenes, L. J. Beaumont, Y. C. Collingham, B. F. Erasmus, M. F. De Siqueira, A. Grainger, L. Hannah, L. Hughes, B. Huntley, A. S. Van Jaarsveld, G. F. Midgley, L. Miles, M. A. Ortega-Huerta, A. T. Peterson, O. L. Phillips, and S. E. Williams. "Extinction Risk from Climate Change." *Nature* 427 (2004): 145–148.

Toynbee, J. M. C. *Animals in Roman Life and Art.* Ithaca, NY: Cornell University Press, 1973.

Uglow, J. *The Lunar Men: The Friends Who Made the Future, 1730–1810.* London: Faber and Faber, 2002.

Walters, M. *A Concise History of Ornithology.* London: Christopher Helm, 2003.

Welty, J. C., and L. Baptista. *The Life of Birds.* 4th ed. New York: Saunders, 1988.

Wheye, D. "Science, Art, and Science Art: Bird Images in the Information Age." *Birding* (October 2004): 524–528.

Wilkinson, R. H. *Reading Egyptian Art: A Hieroglyphic Guide to Ancient Egyptian Painting and Sculpture.* London: Thames and Hudson, 1992.

Acknowledgments

This book is the third phase of a long-term project to raise the visibility of Science Art through images featuring birds. The first two phases are described in the Introduction. All three are the beneficiary of many individuals but would not have been possible without the support and counsel of Paul Ehrlich, whose insights into avian and human ecology are evident throughout. And all three intertwine to some extent, so many of those who offered help or advice in one phase contributed to another. We would like to acknowledge that help and advice here.

Assistance acquiring source material was provided at the Falconer Library, Stanford University, by Michael Newman, Jill Otto, and their support staff, including Courtney Welborn and Becky Horton, and at Stanford's Art and Architecture Library by Alex Ross, Peter Blank, and their support staff, including Katie Keller, Liz Johnson, Vanessa Kam, David Platt, Suzanne Maguire-Negus, Minoti Pakrisi, Linda Treffinger, and Arturo Villaseñor. We extend special thanks to Michael Marrinan for guidance in the art history literature and to Joanna Dyla for guidance through the procedures used in the MARC (Machine Readable Cataloging) Unit and the procedures followed by the Library of Congress in adopting new subject categories.

We relied on the advice of Jared Diamond, Stuart Pimm, Robert Ricklefs, Steve Rottenborn, Joan Roughgarden, and David Wilcove in several of the science-related aspects of the project, on the advice of the gallery directors Mario Bravo, Muldoon Elder, Maria Hiajic, and especially Kathy Foley for gallery- and museum-related aspects, and on the advice of the artists Robert Bateman, John O'Neill, Edward Rooks, and especially Tony Angell, H. Douglas Pratt, Thomas Quinn, and Julie Zickefoose. Cindy Smith, Ted Floyd, George Archibald, Jim Harris, and David J. Wagner provided useful information about journals, organizations, and art history, and John Kriewall and Bill Gomez provided insight into the relationships among science, art, and sponsors. Robert Peck and Doug Wechsler provided background information on exhibitions, installations, and bird photography archives.

The early phases of the project benefited from consultation with Michael Barish, Bill Carver, Bob Lloyd, Nelson Johnson, Charles Kerns, and Tony Russo. Special thanks are due Carel Pieter Brest van Kempen, Rick Wright, and Thomas Grey, whose interest and involvement in bird painting and photography provided new and exciting ways of looking at the subject. We thank Bob Josse, Michael Keller, Channing Robertson, Jill Otto, Carol Killian, Barry Pearlman, Chuck Perry, Michael Gerson, Jef Friedel, Stacy Zapko, and DeDe Evans for facilitating our inquiry into the use of this kind of art in public spaces at Stanford University.

Technical assistance through the Stanford Libraries was provided first and foremost by Makoto Tsuchitani. We also thank Lois Brooks, Barbara Maliska, Sha Xin Wei, Michael Winnick, Victor Haseman, Steve McGriff, Donna Carter, Kim Hayworth, Chris Sanders, Chang Zhao, Jacqueline Mai, Natalie Calvert, Vikram Mohan, Udara Fernando, Alan Hebert, Dana Sheikholeslami, Paul Davis, Kenneth Chan, and especially Deborah Zimmerman. All are great teachers, and their patience is truly remarkable.

Iris Brest and Alice Ohgi provided guidance through the hosting protocols of the university's Web site, and Jim Day provided guidance during the preparation of key graphics. This work was supported in part by funds made available to Stanford's Center for Environmental Science and Policy by Peter Bing, a 1996 Stanford Teaching and Technology Grant, and a 2007 Sloan Foundation Grant made through the foundation's Public Understanding of Science and Technology Program.

With the understanding that we take responsibility for any errors in this book, we thank the following people who read portions of it, enriching it, catching flaws we missed, and asking for clarity when it was lacking: Jim Holland, Pamela Meadowcroft, Laurie Di Battista, and four anonymous readers. And we thank the artists and institutions who allowed inclusion of their images.

We also thank Holly Brady, Lori McVay, and especially Janet Weitz for their support and effort; Jean Thomson Black, who seemed to understand immediately the aim of this interdisciplinary book and found a way to bring the project to fruition; Matthew Laird, the remarkable Mary Pasti, and their associates at Yale University Press; and Russell Galen, who led us to Jean.

Finally, we thank Ida Wheye, who has supported this effort, at personal sacrifice, for as long as we can remember. We know it is not a fair trade, but this book is for you.

Illustration Credits

Frontispiece. Royal Picture Gallery Mauritshuis, The Hague.

Mallee Fowl in Foreword. In John Gould, *The Birds of Australia: In Seven Volumes* (1840–1869), *Leipoa ocellata:* Gould, p. 166. nla.aus-f4773-5-s166. RARE RBN ef F4773. National Library of Australia.

KEO images in Introduction. © www.keo.org.

Plate 1. © 2007 Darryl Wheye.

Plate 2. © Photo: J. Clottes. In J. Clottes, ed., *Chauvet Cave: The Art of Earliest Times* (Salt Lake City: University of Utah Press, 2003). Science Art—Birds

Plate 3. © 1997/2007 Darryl Wheye. Photo courtesy of J. Clottes. Science Art—Birds.

Plates 4 and 5. Photo credit: Alfredo Dagli Orti. © The Art Archive / Corbis.

Plate 6. The Metropolitan Museum of Art, Rogers Fund, 1934 (34.2.1). Photograph © 1993 The Metropolitan Museum of Art.

Plate 7. © Kunsthistorisches Museum, Vienna, Wien oder KHM, Wien.

Plate 8. © Lady Lever Art Gallery, National Museums Liverpool.

Plate 9. Whitney Museum of American Art, New York; Gift of Mr. and Mrs. Benno C. Schmidt in memory of Mr. Josiah Marvel, first owner of this picture, 77.91.

Plate 10. Rock engraving.

Plate 11. © 2005/2007 Darryl Wheye. Science Art—Birds.

Plate 12. Photo credit: George Hughes. Courtesy of the Oriental Institute of the University of Chicago.

Plates 13 and 13a. © The Trustees of the British Museum.

Plate 14. © 1990/2007 Darryl Wheye. Science Art—Birds.

Plate 15. Photo courtesy of Herzog Anton Ulrich-Museum, Braunschweig. Photo credit: Bernd-Peter Keiser. © Herzog Anton Ulrich-Museum, Braunschweig. Kunstmuseum des Landes Niedersachsen.

Plate 16. © Sally M. Berner. Science Art—Birds.

Plate 17. © Photo: J. Clottes. In J. Clottes, J. Courtin, and L. Vanrell, *Cosquer Redécouvert* (Paris: Le Seuil, 2005). Science Art—Birds.

Plate 18. © 2005/2007 Darryl Wheye. Photo courtesy of J. Clottes. Science Art—Birds.

Plate 19. © The Palace Museum, Beijing.

Plate 20. © The Pepys Library, Magdalene College, Cambridge.

Plate 21. © Museo dell' Opera del Duomo, Florence, Italy / The Bridgeman Art Library.

Plate 22. Photo: François Jay. © Musée des Beaux-Arts de Dijon, Dijon, France.

Plate 23. © L. Stroncek. Photo courtesy of the artist. Science Art—Birds.

Plate 24. © The Field Museum, #CSA62002.

Plate 25. Photo credit: Max Hirmer. © Hirmer Fotoarchiv München.

Plate 26. © 1990/2007 Darryl Wheye. Science Art—Birds.

Plate 27. © The British Library Board. All Rights Reserved (Harley 4751 f.36)

Plate 28. © 1996/2007 Darryl Wheye. Science Art—Birds.

Plate 29. © Alan Jacobs Gallery, London, UK / The Bridgeman Art Library.

Plate 30. Photo credit: Banque d'Images. ADAGP / Art Resource, NY. © 2007 C. Herscovici, Brussels / Artists Rights Society (ARS), New York.

Plate 31. © Caves of Lascaux, Dordogne, France / The Bridgeman Art Library.

Plate 32. © 1997/2007 Darryl Wheye. Photo courtesy of Caves of Lascaux, Dordogne, France / The Bridgeman Art Library.   Science Art—Birds.

Plate 33. © Brooklyn Museum, Brooklyn, NY. 49.48, Charles Edwin Wilbour Fund.

Plate 34. © Kunsthistorisches Museum, Vienna, Wien oder KHM, Wien.

Plates 35 and 35a. Reproduced from H. Bub, *Bird Trapping and Bird Banding: A Handbook for Trapping Methods All Over the World,* trans. Frances Hamerstrom and Karin Wuertz-Schafer (Ithaca, NY: Cornell University Press, 1991). Used by courtesy of Cornell University Press.

Plate 36. © Robert Bateman. Reproduction rights courtesy of Robert M. Bateman, Boshkung, Inc. Photo courtesy of Mill Pond Press, Venice, FL.   Science Art—Birds.

Plate 37 © 2007 Darryl Wheye.

Plate 38. © 2005/2007 Darryl Wheye.   Science Art—Birds.

Plate 39. © Georgia Lee.   Science Art—Birds.

Plate 40. © 2004/2007 Darryl Wheye.   Science Art—Birds.

Plate 41. The Museum of Modern Art, New York, NY. Mrs. John D. Rockefeller Jr. Purchase Fund. Photo credit: Digital Image © The Museum of Modern Art / Licensed by SCALA / Art Resource, NY. © 2007 Artists Rights Society (ARS), New York / VG Bild-Kunst, Bonn.

Plate 42. © 1995/2007 Darryl Wheye.   Science Art—Birds.

Plate 43. © Photo: Simon Hazelgrove. Photographic Department, IMSS, Florence.

Plate 44. Presented by Edward Tyrrell, 1863. © The National Gallery, London.

Plate 45. © 2007 The M. C. Escher Company-Holland. All rights reserved. www.mcescher.com

Plate 46. From the picture album *Nokuhitsu gajo.* © Katsushika Hokusai Museum.

Plates 47 and 48. © Tony Angell.   Science Art—Birds.

Plate 49. Printed by permission of the Norman Rockwell Family Agency. © 1967 The Norman Rockwell Family Entities. Photo: The Norman Rockwell Art Collection Trust, Norman Rockwell Museum, Stockbridge, MA.

Plate 50. © The Academy of Natural Sciences, Ewell Sale Stewart Library.

Plate 51. © Henry Bismuth.   Science Art—Birds.

Plate 52. © Paula G. Waterman.   Science Art—Birds.

Plate 53. © Carl Brenders. By arrangement with Mill Pond Press, Inc., Venice, Florida 34285.   Science Art—Birds.

Plate 54. © Roger Tory Peterson.   Science Art—Birds.

Plate 55. © Patricia Pépin.   Science Art—Birds.

Plate 56. © Ray Harris Ching.   Science Art—Birds.

Plate 57. © Thomas Quinn.   Science Art—Birds.

Plate 58. © Tom Grey.   Science Art—Birds.

Plate 59. © Chris Bacon.   Science Art—Birds.

Plate 60. Copyright © Charley Harper.   Science Art—Birds. Published in the *Ford Times* magazine (November 1957).

Plate 61. © Carel P. Brest van Kempen.   Science Art—Birds.

Plates 62 and 63. Used by permission of the Cornell Laboratory of Ornithology.

Plate 64. © 2004/2007 Darryl Wheye.   Science Art—Birds.

Plate 65. © 2005 Darryl Wheye.   Science Art—Birds.

Plate 66. The Metropolitan Museum of Art. Gift of Erwin Davis, 1889 (89.21.3). Photograph © 1993 The Metropolitan Museum of Art.

Plate 67. Full cover of *Science* 308, no. 5727 (June 3, 2005): 1361–1500 [Watercolor by George Miksch Sutton, courtesy of Cornell Laboratory of Ornithology]. Reprinted with permission from AAAS and the Cornell Laboratory of Ornithology. The painting, owned by the Cornell Laboratory of Ornithology, resides in Kroch Library, Cornell University.

Plate 68. Photo: The National Museum of Fine Arts, Stockholm.

Plate 69. © Vadim Gorbatov.   Science Art—Birds.

Index

Page numbers in **bold** refer to illustrations.

Abbot, John, 139
Aeschylus, 45
Aesthetics, xviii, 63–66, 70, 89, 95, 190
Africa, 13, 19, 57, 77; Libya, xxii, **18**. *See also* Egypt, ancient; Nile River
Afterlife, 23, 57
Alaska: ducks, 72; owls, 43
Albertus Magnus, 47, 49
Aldrovandi, Ulisse, xxiv, **46**, 47
Allen, Arthur, 132, 139
Angell, Tony, x, xxv–xxvi, **88**, 89
Animals, 53, 55, 57, 66–67, 132; art, 66, 121; and humans, xxvi, 107. *See also* Birds; Mammals; Reptiles
Animism, 67
Archaeopteryx, xxiii, 35
Aristotle, 57, 16
Art, vii, 40, 59, 64, 127; animal, 66, 121; archive of, xvi, 53, 97; Chinese art, 33, 67; and conservation, 53, 59, 143; content of, 67, 95; cultural evolution and, 43, 70, 85; Dutch, 52, 127, 137; Egyptian, xxi–xxiv, 5, 9, 16, 52, 55, 57, 67, 118; elements and conventions of, 67, 69, 77, 87; evolutionary view of, 43; Flemish, 137; Greek, xxii, 16, 23, 45, 67, 137; graphic databases, 132; Information Age, 1; medieval, 1, 33, 66, 67; nature-oriented, xiii, xv, 25, 65, 107, 141; in prehistory, 1; Renaissance, 1, 52, 66–67; rock, 19, 75, 77; Roman, xxii, xxiii, 16, 25, 39, 51, 67, 137. *See also* Aesthetics; Bird art; Illustrations; Paleolithic cave art; Photographs; Science Art
Art interpretation, xvii, 33, 55, 66, 85, 89; through lens of culture, 137; through lens of science, xviii, 65, 87
Art styles, 66–68, 95, 137; baroque, 52; expressionism, 79; impressionism, xix, 27, 67, 137; minimal realism, xxvii, 117; painterly, xxvi, 95, 111; photo-realism, 27, 95–96, 107–109; realism, 68, 111–115; representational, 121, 137; rococo, 25; surrealism, 51. *See also* Photographs
Art techniques: scientific perspective, 67, 77, 85; *sfumato*, 67
Artist Registry for Ornithological Researchers, xvii, 169
Assyrians, 19, 25
Athena/Minerva, 72
Attar, Farid ud-Din, 49
Audubon, John James, ix, xxvii, 53, 61, 97, **122**, 123, 133
Auks, 66–67; Great Auk (*Pinguinus impennis*), xxii, 21, 31, 123

Automobiles, xxvi, 107
Avian influenza, xxii, 2, 27. *See also* Virus
Avocet (*Recurvirostra avosetta*), 109

Babylon, xxiii
Bacon, Chris, xxvii, **114**, 115
Bacon, Roger, 35, 81
Banding and monitoring birds, xxiv, **60**, 61. *See also* Bird watching
Barnacles: Goose-necked Barnacle (*Lepas* sp.), xxiv, **46**, 47; stalked, 47
Bartsch, Paul, 61
Bateman, Robert, xxv, **62**, 63, 93
Beebe, William, 35
Behavior, xvi, 101, 103. *See also* Bird behavior
Belgium, 49, 51, 59; Flemish art, 137
Bennett, Thomas, 139
Berner, Sally M., xxii, **26**, 27, 113
Bestiaries, 1
Bible and birds, 16, 19
Binoculars, x, 52, 143; field glasses, 52, 133; monoculars, 52
Biogeography, 105, 109
Biology: bird, xvi, xvii, 61; development of, 66; recorded by artists, 40
Biomimicry, 129
Biophilia, 132
Bird art, xxiii, 33, 40, 52–53; chimeras as, xxv, 5, 43, 51, 72, 75; and decoding nature, 40; masks as, 5, 55, 67; and technology, 79; Web site, xvii. *See also* Nature, images of; Paleolithic cave art
Bird artists, xvii, xix, 33, 133, 167–169
Bird banding and monitoring, xxiv, **60**, 61. *See also* Bird watching
Bird behavior, xxvi, 49; defending territory, xxvii, 107, 141; displaying, 40–41; fighting, 16; flocking, 49, 105, 135, 141; foraging, 13, 41, 49, 51, 55, 89, 105, 111; mobbing, x, 17, 41, 103; perching, 55, 77, 79; preening, 21, 40; roosting, 13, 133, 141; sleeping, 70; thermoregulating, 40, 70, 117. *See also* Bird mating and breeding; Commensalism
Bird conservation, xxiv, 39, 52–53, 59, 139, 143; and feeders, 105; invasive species and, 41, 109; in New Guinea, 55; and projected extinctions, 39. *See also* Conservation
Bird mating and breeding: cooperative breeding, 19; courtship and mate selection, xxiv, xxvii, 21, 49, 141; cuckoldry, 70; grounds and range, 33, 47, 61, 105, 141;

Bird mating and breeding (continued): leks, xix, xxvii, 49, 113, 141; mating systems, 49, 111; megapods, xi; nesting, xxii, xxvii, 19, 43, 59, 75, 129; parents and young, xi, 63, 103, 111; season, xii, 21; territory, 103

Bird physiology, 19, 51, 89; bills and mandibles, xxv, 79, 89, 107, 113, 115; coloration, 40; fat, 117; feathers, xxii, 2, 16, 35, 40, 61; phenotype, 69; preen oil, 40, 57; and respiratory rate, 83. *See also* Biology

Bird vocalizations, 49, 99, 139; calls, 55, 77; notation of, xxv, 40, 61, 79; song, xi–xii, 40, 79, 91, 99

Bird watching, xix, 75, 91; bird counts, 105; and feeder studies, 105; and feeding, xxvi, 95

Birdlime, 17

Birds, xii; and Bible, 16, 19; biogeography, 105; biology, xvi, xvii, 61; biomimicry, 129; changes in geographic range, 39, 105, 107; coevolution, 51; domesticated, xxii, 16, 21; ecology, 40, 63, 103; exotics, 16, 53, 137; evolution, xxiii, 35, 41, 49, 87; flight, xxiii, 35, 39–40, 77, 81, 113; migration, xxiii, 21, 33, 39–40, 61, 105, 113; names, 47, 49; navigation and orientation, 25, 40, 61; predators, 52, 89, 103, 111, 115, 121, 141; prey, 23, 45, 49, 87, 123, 141; sea and shore birds, 72, 115; songbirds, 5, 16, 37, 83, 99; specimen collections, xxii, 17, 49; urban, 121; wild, xxii, 37, 52, 115; waterfowl, 21. *See also* Bird art; Birds, value of, to humans; Mammals; Ornithology; Reptiles

Birds, value of, to humans, xvi–xvii, xix, 16, 139; as captives or pets, 52–53, 57, 127, 131, 137; as companions, xix, 2; as conservation icons, 63, 139; in decoding nature, 40; on Easter Island, 75; falconry, ii, x, 2, 16, **22**, 23, 61; as food, xxii, 21, 51; for implements in the Paleolithic, 17, 55, 59; as indicator species during the Paleolithic, 55; as intermediaries, xxi, 5; as models for human behavior, 5; as models for replicating flight, 2, 81; for pest control, 57, 93; as providers of free services, 16; as symbols of luck; 53; in testing toxic gases, 37. *See also* Humans and animals; Hunters

Birds in Art exhibition (Woodson Museum), xix, 27

Bismuth, Henry, xxvi, **98**, 99

Blackbirds, xxii, xxv–xxvi, 16; Brewer's Blackbird (*Euphagus cyanocephalus*), 93

Bluebirds, 72

Bonde, Neils, 35

Bonney, Rick, 105

Boran people of Kenya, xxv, 77

Boucher, François, xxii, 16, **24**, 25

Boulton, Matthew, 83

Boyle, Alison, 85

Boynton, Mary Fuertes, 16, 93

Brasher, Rex, 139

Brenders, Carl, x, xxvi, **102**, 103, 139

Brest van Kempen, Carel Pieter, xxvii, **120**, 121, 167

Breuil, L'Abbé Henri, 43

Brower, Kenneth, 127

Brueghel the Elder, Pieter xxiv, **58**, 59

Bub, Hans, 61

Burkina Faso, 5

Bushtit (*Psaltriparus minimus*), xxvii, **128**, 129

Bustard, 55

Caesar, Julius, 25

Cameras, vii, 97, 113, 133, 141. *See also* Photographs

Canary in coal mine, xi, xxiii, 2, 37

Captions, x, xiii, xv, xvii–xix; contents of, 70, 118, 121; functions of, 68–70, 118, 123. *See also* Science Art

Cardinal, Northern (*Cardinalis cardinalis*), 105

Carson, Rachel, xi

Caves with Paleolithic bird art: Chauvet, x, xv–xvii, **xvi**, **xvii**, xxi, 1, 7, 43, 67, 91; Cosquer, xxii, **31**, 66–67; and lamps, 31; Lascaux, xxiv, 7, 31, 52, 55, 67; Les Trois Frères, xxiii, 7, **43**, 67. *See also* Paleolithic cave art

Chaucer, Geoffrey, xxiv, 49

Chauvet Cave, x, xv–xvii, **xvi**, **xvii**, xxi, 1, 7, 43, 67, 91

Chefren (king), 1, **8**, 9, 55, 118

Chickens, 16, 21

Chimeras as bird art, xxv, 5, 43, 51, 72, 75

Chinese art, 33, 67

Ching, Ray Harris, xxvi, 97, **108**, 109

Christiansen, Per, 35

Cicero, 16, 49

Cinteotl (deity), xxi

Clarke, T. H., 123

Climate, 1, 73; change, effects of, xii, 19, 31, 33, 61, 105; change, global, xi, xxii, xxvi, 39, 49, 105; Medieval Warm Period, 49

Clines, Francis X., 75

Cochoa, 33

Cockatoos, 37, 53; Sulphur-crested Cockatoo (*Calyptorhynchus galerita*), 83

Codex Laud, xxi

Coevolution, xxiv, 51, 77, 99. *See also* Evolution

Coheleach, Guy, 139

Coins, ancient Greek, xxiii, **44**, 45

Coloration, cryptic, 40

Colson (Jean-François Giles), xxiii, **36**, 37

Columbus, Christopher, 53

Commensalism, xxvi, 55, 95, 101, 111

Commodus (emperor), 19

Communication, 65

Computers, xvii, 61; art databases and, 169. *See also* Information Age, Science Art

Conservation, xii, xvi–xvii, 53, 63, 109, 139; and birds, 53, 59, 143. *See also* Art: nature-oriented; Bird conservation

Constable, John, 65

Conway, Bevil, 96

Cormorant, Japanese (*Phalacrocorax capillatus*), 21

Cornell University, 132, 134, 139; Cornell Laboratory of Ornithology, 105

Corvids, 99

Cosquer, Henri, 31

Cosquer Cave, xxii, **31**, 66–67

Courbet, Gustave, 137

Cranes, 21, 33, 53, 55; Crane (*Grus grus*), 33

Crowberries, 117

Crows, 7, 99; Carrion Crow (*Corvus corone*), 15

Cuckoos, 33

Curlews, 9; Eskimo Curlew (*Numenius borealis*), 53, 117; Long-billed Curlew (*N. americanus*), 113, 115

Daedalus, xxiii, 2, 29, 35, 81

Damiani, Pietro, 47

Darwin, Charles, 83

Darwin, Erasmus, 83

De Hamilton, Carl Wilhelm, xxiv, **48**, 49

Degas, Edgar, 137

Diamond, Jared, 55, 75
Diclofenac, 127
Diderot, Denis, 25, 66
Diseases, bird-related, xxii, 2, 15, 27, 99
Dodo (*Raphus* sp.), 1, 53
Doughbirds, 117
Dovecotes, 51
Doves: as icon, xxi, 1, 11; Rock Dove (*Columba livia*), 25. *See also* Pigeons
Dragons, 5, 51. *See also* Chimeras as bird art
Drugs, veterinary, and wildlife losses, 127
Ducks, x, 16, 21, 33, 52, 55, 129
Dürer, Albrecht, 123
Dutch art, 52, 127, 137

Eagles, xxiii–xxiv, 1, 5, 33, 45, 55; Bald Eagle (*Haliaeetus leucocephalus*), 41; Golden Eagle (*Aquila chrysaetos*), 55, 103. *See also* Mobbing
Easter Island, 72, 75; Motu Nui, 75
Ebers, George, 57
Eckelberry, Don, 132, 134, 139
Ecology, xii, xv–xvi, 39, 45, 63, 87, 105, 107; bird, 40, 63, 103; edge effect, 141; movement, 15, 61; niche specialization, 134
Education, environmental, 127. *See also* Science Art: and education
Edwards, David, 64
Egrets, xxiv; Cattle Egret (*Bubulcas ibis*), 55
Egypt, ancient: art, xxi–xxiv, 5, 9, 16, 55, 57, 67, 118; animals, 52, 53, 57; birds, xxiv, 5, 19, 21, 25, 55, 57; conservation message in art, 52; Giza, 9; medicine in, 57; Saqqara, 57; specific bird names, 1; Thebes, 21; Tuna el Gebel, **56**, 57. *See also* Nile River
Ehrenstrahl, David Klöcker, xix, xxvii, **140**, 141
Ehrlich, Paul, vii, xvii, 51
Einstein, Albert, 65
Elephant Bird, Great (*Aepyornis maximus*), 19
England: art compared to Chinese art, 33; Sutton Hoo, **22**, 23
Environment, 61, 63, 107, 131. *See also* Ecology
Environmental Literacy Council, 131
Environmental Movement, xi, 132
Environmental Science Testbank, 131
Escher, M. C., xxv, 66, **84**, 85
Ethelbert II, 23
Ethofer, Theodor Josef, 123
Evolution, xvi, 43, 51, 129, 143; bird, 41, 49, 87; cultural, 43, 70, 85; and flight, xxiii, 35. *See also* Coevolution
Exotics, 16, 53, 137
Exploration, Age of, 17, 143
Extinction, 19, 63, 132

Fabritius, Carel, 137
Falconry, ii, x, 2, 16, **22**, 23, 61
Falcons, 1, 11, 121; Barbary Falcon (*Falco pelegrinoides*), 1; in Egypt, 9; Eleonora's Falcon (*F. eleonorae*), 1; Eurasian Hobby (*F. subbuteo*), 1; as Horus xxi, xxiii, **8**, 9; Lanner Falcon (*F. biarmicus*), ii, 1; Merlin (*F. columbarius*), 1; Peregrine Falcon (*F. peregrinus*), 1, 25; Red-footed Falcon (*F. vespertinus*), 1; Saker Falcon (*F. cherrug*), 1; Sooty Falcon (*F. concolor*), 1
Faulkner, William, 135
Feeding birds. *See* Bird watching

Ferdinand, Duke, 61
Field glasses. *See* Binoculars
Field guides, viii–ix, xxiv, 49, 91, 105, 134, 143
Finch, Otto, 137
Finches, 33; Darwin's, 2; House Finch (*Carpodacus purpureus*), 105
Fish, 7, 21, 41, 47, 51
Flemish art, 137
Flicker, Northern (*Colaptes auratus*), 141
Flight: bird, 35, 39–40, 77, 81, 113; birds as models for replicating, 2, 81; mechanical, xxiii, 35; ultralight, xxv, 8
Flu, bird, xxii, 2, 27. *See also* Virus
Flycatchers: Old World, 33; Peewee, 61
Forests, 63, 134
France: Cape Morgiou, 31; Marseilles, 31; Pyrenee Mountains, 43. *See also* Caves with Paleolithic bird art
Franklin, Benjamin, 27, 83
Frederick II (Holy Roman Emperor), 49
Fuertes, Louis Agassiz, ix, xxvi, 16, **92**, 93

Galison, Peter, 85
Ganymede, 5
Gay, John, xxi, 13
Geese, 16, 21, 39, 40; Barnacle Goose (*Branta leucopsis*), 47; Canada Goose (*B. canadensis*), 39; Snow Goose (*Chen caerulescens*), 39
Genes, 51
Genghis Khan and pigeon mail, 25
Gherardini, Lisa, 96
Glacial Maximum, 31
Goldfinch (*Carduelis carduelis*) and plague, 5
Gombrich, Sir Ernst, xviii, 66, 67
Google search for Science Art, 125
Gorbatov, Vadim, xix, xxvii, 113, **140**, 141
Goshawk, Northern (*Accipiter gentilis*), 141
Gould, John, ix–xi, **xii**
Gould, Stephen J., 43
Gounod, Charles-François, 91
Goya, Francisco, 137
Greece (ancient), art of: xxii, 67; coins, 45; falconry, 23; pet birds, 16, 137
Gregory, Saint, **10**, 11
Grey, Tom, xxvi, **112**, 113
Griffins, 5, 51. *See also* Chimeras as bird art
Grosbeak, Evening (*Coccothraustes vespertinus*), 95, 105, 107
Grouse, xxvii, 49, 55; Black Grouse (*Tetrao tetrix*), 131, 141
Guineafowl, 109
Gulls, 33, 103
Guthrie, R. Dale, 43
Gyrfalcon (*Falco rusticolus*), ii

Habitats, xii, 47, 57, 61, 63, 139
Handel, Frederic, 91
Harley Manuscript, **46**, 47
Harper, Charley, xxvii, **116**, 117
Hartlaub, Gustav, 137
Hawks, 1, 33, 53, 61, 103
Heinrich, Bernd, 99
Henry IV, 61

Herodotus: and animals in Egypt, 52–53; and disease in Egypt, 57; and Ostrich, 18, 19
Herons: in ancient Egypt, 5; as decoys, 21; Gray Heron (*Ardea cinerea*), xxiv, 61; Great Blue Heron (*A. herodias*), 60, 61
Hieroglyphics, 1, 9, 21, 118
Hobby, Eurasian (*Falco subbuteo*), 1
Hoffmann, Hans, 123
Hogarth, William, 16
Hokusai (Katsushika Hokusai), 52, 87
Holbein the Younger, Hans, ii
Holocene epoch and sea level, 31
Honeyguide, African Greater (*Indicator indicator*), xxv, 76, 77
Hoopoe (*Upupa epops*), 49
Hoose, Phillip, 132
Horemheb, 20, 21, 118
Hornbills, 121
Horus, xxi, xxiii, 8, 9
Huang Quan, 32, 33
Hugh of Fouilloy, 11
Humans and animals, xxvi; interactions, 107. *See also* Birds, value of, to humans
Hummingbirds, xxvi; coevolution with flowers, xxiv; Rufous Hummingbird (*Selasphorus rufus*), 111
Humphrey, Philip, 91
Hunters: as artists, 66; disguised, 43; use of fletched feathers, 43; in Lascaux Cave, 55; in symbiotic relationship, 77. *See also* Falconry; Humans and animals

Ibis: in ancient Egypt, 53; Sacred Ibis (*Threskiornis aethiopicus*), xxiv, 52, 57. *See also* Egypt, ancient
Icarus and Daedalus, xxiii, 29, 35, 81
Icons, xvi–xvii, xxi; biological basis of, 1; birds as, xxi, 1, 5, 11, 63, 137, 139; cartoon as, 132, 142; cherubs as, 5, 25; cultural or religious, 73; of nature, 137; specific plant or animal as, 25, 45, 59; symbol of protection and, 9
Illustrations, vii, xxv; vs. art, xxvi, 89, 91, 93, 95; book plates, 49; diagrams, 89; history of, viii, 91
Imagination, 64, 65, 79, 89
India and vulture deaths, 127
Information Age: art, 1; graphic databases, 132. *See also* Web addresses
Insects: African honeybees, xxv, 77; Alfalfa Weevils (*Hypera postica*), 93; Long-horned Beetles (Cerambycidae), 134; Monarch Butterfly (*Danaus plexippus*), 89; mosquitoes, 99, 111; moths, 7, 121; parasitic flies, ix; spiders, 129
Inspiration and intuition, 64
Inuits and ducks, 72
Italy: Florence, xxiii, 35; Pompeii, 52; Sicily, xxiii

Jackson, Jerome, 139
Jones, Caroline, 85

Kalahari Bushmen, 19, 55
Kalmbach, E. R., 93
Kennedy, Donald, ix, xvii
Kenya, the Boran people of, 77
KEO, xviii, **xviii**; UNESCO and, xviii
Kestrels: Eurasian Kestrel (*Falco tinnunculus*), 1; Lesser Kestrel (*F. naumanni*), 1

Kites in ancient Egypt, 1
Klee, Paul, xxv, 65, 78, 79
Koestler, Arthur, 65
Kuhl, J. J., 139

Larks, 83
Lascaux Cave, xxiv, 7, 31, 52, 55, 67
Lee, Georgia, xxv, 74, 75
Leks, xix, xxvii, 49, 113, 141
Lent, 47
Leonardo da Vinci, 93; replicating flight, xxv, 35, 80, 81
Leroi-Gourhan, André, 7, 43
Les Trois Frères (Three Brothers) Cave, xxiii, 7, 42, 43, 67
Libya, xxii; Neolithic Ostrich image, 18
Light, viii, 66
Literacy, 25
Livestock, 51, 127
Livingston, Margaret, 96
Longrich, Nick, 35
Loons, 129
Lord God Bird, 132, 135
Lorenz, Konrad, 81
Louis XV, 25
Lunar Society, xxv, 13, 83

Ma'at (deity), 19
Macaws, 53
Madagascar, 19
Magpies, 99
Magritte, René, xxiv, 50, 51
Malleefowl (*Leipoa ocellata*), x–xi, **xi**
Mammals: cats, 52, 57; cattle, 55; dogs, 52, 57, 127; gazelles, mummified, 57; hares, 45; horses, 13, 5, 7, 51; lions, 9; mammoths, 7; monkeys, 109; mummified, 57; rabbits, xxiii, 51; rats, 127; rhinoceros, 55, 123; roe deer, 7; sacred cows, 127; squirrels, 109; vision of, 132; wolves, 99, 101
Mammoth Steppe, 67
Manet, Édouard, xix, xxvii, 131, **136**, 137
Masks as bird art, 5, 55, 67
Materia Mecia, 33
McGowan, Kevin, 99
McLuhan, Marshall, 85
McNeill, William, 67
Medawar, Peter, 64
Megapods, xi
Menageries, private, 17
Mereruka, *mastaba* of, 21
Mergansers, 5
Merlin (*Falco columbarius*), 1
Metaphors, 45, 87
Metatags for Web page, 167
Meurend, Victorine, 137
Meyer, Leonard, 65
Michelangelo, 93
Microimages and Homing Pigeons, 25
Microraptor gui, as four-winged, 35
Middle Ages, art of, 1, 33, 66, 67
Migration. *See* Birds: migration
Minerva, 72
Mobbing, x, 17, 41, 103
Mockingbird, Northern (*Mimus polyglottos*), 123

Models, birds as, 33, 43. *See also* Flight
Mona Lisa, scientific analysis of, 66, 96
Monet, Claude, 137
Monk's drawing book, **32**, 33, 123
Morris, Arthur, 169
Mortensen, Hans, 61
Mozart, Wolfgang Amadeus, 91
Musée de l'Homme, 7
Museum Wormianum, 123
Museums: Copenhagen Museum, 123; Leigh Yawkey Woodson Art Museum, xix, 27, 115; Louvre, 57; Museum of Modern Art, New York, 15, 79; National Gallery, Berlin, 79; National Gallery, Washington, DC, 123; Smithsonian Institution, xvii, 61, 91
Music, 91; meaning and communication, 65

Napoléon, 25
Narratives: of biological organization, 69; of culture, 137; and route through image, 77
National Academy of Science, 64
National Audubon Society, 75, 91, 105, 132, 135
National Endowment for the Arts, 168
National Science Foundation, 168
Natural history, 21, 47
Natural painture, 65
Naturalism, Age of, 132
Nature, images of, xiii, xv, 26, 65, 107, 141; birds, xv; decoding, 40; painting from, 25; perception of, 70. *See also* Art; Conservation; Environment; Evolution; Habitats
Nature Conservancy, 139
Neckam, Alexander, 47
Nectanebo II, **8**, 9, 55
Nest, x, xi, xxiii, xxvii; architecture, xviii; robber, xxiv. *See also* Bird mating and breeding
New Guinea, 5, 55
New Zealand, 19
Nightingales, 33; Nightingale (*Luscinia magarhynochos*), 40
Nightjars, ix, 33; Common Nighthawk (*Chordeiles minor*), 121
Nile River, 5, 9, 21, 25, 57
Nobel Prize, 64, 73
Nomenclature, binomial, 143

Oberholser, Harry Church, 93
Okri, Ben, 73
Omnivores, 101
Opportunism, 121
Optics, 52, 70, 133. *See also* Binoculars
Oriole, Black-backed (*Icterus abeille*), 89
Ornithology, xix, 49, 75, 131, 137, 143
Osprey (*Pandion haliaetus*), 41, 59
Ostriches (*Struthio camelus*), xxii, **18**, 19, 55
Owls, xxiii, 1, 40, 43, 63, 72; Barred Owl (*Strix varia*), 63; Eagle Owl (*Bubo bubo*), 7; Great Horned Owl (*B. virginianus*), 111; Long-eared Owl (*Asio otus*), 7; Snowy Owl (*Nyctea scandiaca*), bones of, 43; Spotted Owl (*S. occidentalis*), 63

Pakistan and vulture deaths, 127
Paleolithic cave art, xviii, xxi, 31, 40, 43, 66–67, 91; artists and, 31, 43, 81; birds and animals portrayed in, 7, 31, 43, 55; classification of, 43; conservation message in, 52; determining date of, 31, 43; interpreting, 7, 43, 55; model of parental care in, 43; owl bones and, 43; owl-headed human figure in, 43; relationships portrayed in, 43, 55, 67; survey of, 7, 55. *See also* Caves with Paleolithic bird art
Papyrus, 9, 57; Ebers, 57
Parrots, 16, 53; African Grey Parrot (*Psittacus erithacus*), 137
Partridges, 55
Passerines, 16, 83
Peacocks, 16
Pelicans, 16, **20**, 21, 118; Dalmatian Pelican (*Pelecanus crispus*), 21; Pink-backed Pelican (*P. refescens*), 21; White Pelican (*P. onocrotalus*), 21
Penguins: body shape, 40, 129; "Northern Penguin," 31
Pennycuick, Colin, 13
Penrose, Roger, and the Penrose Staircase, 85
Pépin, Patricia, **106**, 107
Persia, 19
Persian Gulf, 57
Peter the Deacon, 11
Peterson, Roger Tory, 91, 95, **104**, 105, 141
Pheasants, 33
Philosopher Giving a Lecture on the Orrey, A (Wright), 83
Phoebe (Peewee Flycatcher), 61
Photographs, vii–viii, xxvii, 97; as aid to artists, 141; archive of, 97; as art, 121; digiscoping, 97; vs. minimal realism and photo-realism, 117, vs. painting, 115, vs. realism, 113, vs. representational art, 137; stop-action, xix, 141; strobe-light illumination, 41
Photo-realism. *See* Art styles
Pictographs, xxii, 55
Pigeonholes, 51
Pigeons, xxii, 16, 21, 25, 51–53; Homing Pigeon (*Columba livia*), 2, 16, 25; and mail network, 25; Passenger Pigeon (*Ectopistes migratorius*), 97
Pintail (*Anas acuta*), 61
Pisano, Andrea, xxiii, **34**, 35
Planck, Max, 65
Plants, seed-producing, xxiv, 51
Plautus, 13
Pliny, 19, 57
Plotinus, 65
Poincaré, Henri, 64
Polymaths, 70, 81
Polynesia, 72
Predators and prey, x, xxvii, 45, 55, 73, 87. *See also* Birds: predators; Birds: prey
Prey. *See* Predators and prey
Priam, 45
Priestly, Joseph, 83
Priests, 21, 53
Prometheus, 5, 45
Protein, 47, 51, 55
Public spaces. *See* Venues
Puffin (*Fratercula arctica*), 31

Quails, 16, 21
Quan Huang. *See* Huang Quan
Quinn, Thomas, xxvi, 97, **110**, 111

Rabies, 127
Raptors, 1, 23, 33, 40, 55, 57, 141
Raven, Peter, 51
Ravens, xxi, xxiv, xxvi, 13, 55, 99, 101, 103; Common Raven (*Corvus corax*), 40, 95, 99, 101, 103; corvids, 99
Razorbill (*Alca torda*), 31
Realism. *See* Art styles
Rembrandt van Rijn, 96
Renaissance art, 1, 52, 66–67
Reptiles: dinosaurs, feathered, 5; lizards, 87; snakes, 57, 87, 123
Resources, mismanagement of, 75
Reynolds, Sir Joshua, 137
Richard I (the Lion-Hearted), 47
Richardson, John Adkins, 79
Rigor Vitae: Life Unyielding (Brest van Kempen), 121
Ripley, S. Dillon, 91
Rockwell, Molly, 91
Rockwell, Norman, xxvi, **90**, 91
Roosters, 109
Root-Bernstein, Robert, 64, 81, 85
Rome (ancient), art of, xxii, 16, 51, 67; and birds, xxiii, 19, 25, 39
Royal Society, 13, 83
Royal Society for the Protection of Birds (RSPB), xvii, 59
Rubens, Peter Paul, 137

Sage, Balthazar, 37
Sargent, John Singer, 133
Science, 65, 67, 85, 87; and art, xvi–xix, 64; experiments, 13, 64, 83; lens of, and art, 65, 87; mathematical laws, 81; and neutrality, 65. *See also* Evolution
Science Art: xvi–xix, 69, 115; accuracy of, 68, 95; and art history, xvii, 21, 67; and artistic license, 95; as category of art, 131; checklist, 167–169; classification of content, 137; and community activities, 169; definition, xiii; vs. impressionism, xix, 137; market for, 16, 53; medium and, 97; records of, 72, 129, 131; requirements of, 69
—and education, 72, 129; fee-free image use, 129; teaching environmental issues, xiii, xvii–xix, 127, 168
—finding examples of, 72, 169; in AMICO, 169; in ARTstor, 169; checklist for labeling, using, finding, xix, 167–169; in collection inventories, 72; copyright lines and, 167; credit lines and, 93; finding artists, 169; finding online, 168; Google search, 125; Web site, xvii
—venues, 27; with community activities, 169; exhibit spaces, 125, 127, 129, 168; and public notices, 125; in public spaces, xxvii, 51
Science Artists: checklist for, 167–169; research available to, 41; support for, 169; Web site, xvii
Science magazine, xix, xxvii, 134, 139
Scientific Revolution, 132; art of, 1
Scientists: and artists, xxv, 85; citizen, 105, 143; term, 83
"Scipio's Dream," 49
Sclater, Philip Lutley, 137
Seabirds, 72
Seeds, xxiv, 51; crop, 93
Senescence, 15, 107
Sense impressions, 67
Shaman, 43, 55, 66

Sharks, 75
Shepard, Paul, 72
Shorebird, 115
Sibley, David, 134
Singer Tract, 132, 134
Sistine Chapel, 93
Skylark (*Alauda arvensis*), 33
Small, William, 83
Smit, Joseph, 137
Snow, C. P., 64, 178
Society of Animal Artists, 121, 169
Solomon, 16
Songbirds, 5, 16, 37, 83, 99
Sotheby's, 123
South Africa, 19
Sparrows, 83, 93; Song Sparrow, 135; Tree Sparrow (*Passer montanus*), 33
Spears, 43
Sphinx, 5, 9, 21, 51
Spoonbill (*Platalea leucorodia*), 33
Squabs, 51
Stanford University, xvii, xxvii, 127, 132; Science Art Web site, xvii
Starling, European (*Sturnus vulgaris*), 61, 91
Stereoblindness, 96
Stevenson, Robert Louis, 93
Stork: in middens, 55; White Stork (*Ciconia ciconia*), 61
Strabismus, divergent, 96
Stroncek, Lee, xxiii, **38**, 39
Struthophangi and ostrich hunting, 19
Stubbs, George, xxi, **12**, 13, 123
Sutton, George Miksch, ix, xix, xxvii, **138**, 139
Sutton Hoo, **22**, 23
Swans, 16, 55
Swifts (*Apus apus*), 81
Synscience, 81

Tang Dynasty, 33
Tanner, James, 134
Taxonomy, xvii
Teal, European (*Anas crecca*), 61
Tenison Psalter, 33
Tern, Sooty (*Sterna fuscata*), xxv, **74**, 75
Territory. *See* Bird behavior; Birds: ecology
Texas: Big Bend, 93
Tharsis, 16
Theron, 45
Thomas, C. D., 39
Thoth, 57
Thrasher, Brown (*Toxostoma refum*), 123
Thrush, 91; Ferruginous, 123
Thunderbird, 5
Titmouse, Tufted (*Parus bicolor*), 105
Tombs, 21, 23
Topography, 73
Toucan, Plate-billed Mountain, 109
Towers of Silence, 127
Troy, 5; Trojans, 45
Turkeys, xxii, 27, 113; banding, **60**, 61; Wild Turkey (*Meleagris gallopavo*), 27
Turner, James, 139
Tuthankhamen, xxiii
Tuthmosis IV, xxii

Unconscious, the, 64, 65
Universities, fourteenth-century, 47
U.S. Bureau of Biological Survey, 93
U.S. Department of the Interior, 139
U.S. Fish and Wildlife Service, 63, 93
U.S. government and commissioning art, 16
U.S. Signal Corps, 25

Varro, 16
Venues, 93, 125, 168; in public spaces, xxvii, 125, 127, 129
Virus: H5N1, xv; West Nile, 2, 15, 99
Vision, human vs. mammal, 132
Von Guericke, Otto, 83
Vultures, 2, 13, 40, 55; deaths in India and Pakistan, 127; Long-billed Vulture (*Gyps indicus*), 127; Oriental White-backed Vulture (*G. bengalensis*), **126**, 127; Ruppell's Griffon Vulture (*G. rueppellii*), 13; Turkey Vulture (*Cathartes aura*), 1; White-backed Vulture (*G. africanus*), 13

Wagtails, 33
Warblers, partitioning by, viii
Wars, 15, 16, 23, 25, 37
Waterfowl, 21
Waterman, Paula, xxvi, **100**, 101
Watt, James, 83
Web addresses: birds.stanford.edu, xvii; scienceart-nature.org, xvii. *See also* Science Art
Wedgwood, Josiah, 13, 83
Wheatear, Northern (*Oenanthe oenanthe*), 33
Wheye, Darryl, ix, xvi, xxi–xxv, xxvii, 6, **14**, 18, 22, **26**, **28**, 30, **38**, **40**, **42**, 44, 46, 55, **64**, **65**, 74, 76, 78, 126, 128
Whimbrels, 115
Whip-poor-will (*Caprimulgus vociferus*), ix
Wick, James, xxv, 81
Wilderness, urban, 52
Wilson, Alexander, 133
Wilson, E. O., vii, 132
Wing of a Blue Roller [*Coracias garrulous*] (Dürer), 123
Winged Migration (film), 97
Woodpecker, Woody, 132
Woodpeckers, xix; Green Woodpecker (*Picus viridis*), 33; Ivory-billed Woodpecker (*Campephilus principalis*), 52, 53, 73, 131–134, 135, 139; Pileated Woodpecker, 135
Wright, Joseph (of Derby), xxv, 37, **82**, 83
Wyeth, Andrew, xxii, **14**, 15
Wyeth, N. C., 93

X-ray analysis of mummified animals, 57
Xu, X., 35

Zeus, 5
Zhang, F., 35
Zhou, Z., 35
Zickefoose, Julie, 139
Zoos, 17; Cincinnati, 97; private menageries, 17

Frontispiece. *Portrait of Robert Cheseman*, 1533
By Hans Holbein the Younger (1497–1543)
Height 23 1/4 × 24 3/5 in. (59 × 62.5 cm)

By convention, falconry, the 4,000-year-old sport of kings, paired social rank with a particular species of bird to train. The king had his Gyrfalcon (*Falco rusticolus*), and the squire—that is Cheseman's apparent rank—had his Lanner Falcon (*F. biarmicus*). The Latin lettering by Holbein in the background identifies the falconer, his age, and the date. Repeatedly, contemporary authors misidentify Cheseman as Henry VIII's personal falconer and misidentify the bird as a Gyrfalcon (the telltale buff-colored feathers on the back of the head distinguish this falcon as a Lanner).

Of the 1,000 or so books on falconry, Emperor Frederick II's *The Art of Falconry* (c. 1248) was the first to record detailed observations and life histories. It also records a natural world rich in wildlife and a society seemingly on the verge of taking closer notice of its birds: "The pursuit of falconry," Frederick II advised, "enables nobles and rulers disturbed and worried by the cares of state to find relief in the pleasures of the chase. The poor, as well as the less noble, by following this avocation may earn some of the necessities of life; and both classes will find in bird life attractive manifestations of the processes of nature. The whole subject of falconry falls within the realm of natural science, for it deals with the nature of bird life."